Valorization of Dredged Sediments as Sustainable Construction Resources

Valorization of Dredged Sediments as Sustainable Construction Resources provides a comprehensive and up-to-date overview of research on the reuse of sediments as a raw material for civil engineering fields. Dredging is increasing in worldwide scale, while deep sea disposal is being eliminated and landfill disposal is becoming less sustainable. Yet these sediments offer a potentially useful resource for production of pavements, bricks, and ceramics, for functional soil and road construction, also, as fine and lightweight aggregates, supplementary cementitious materials, and geopolymer binders and even as raw material for clinker production. This has sparked considerable research on valorization, which is presented here. This book:

- Covers the main re-uses of sediments and the valorization in the field of civil engineering
- Focuses on up-to-date research on practical guidance
- Discusses limitations and future perspectives for sediments valorization

The book suits researchers and students working in the area, as well as industrial engineers and professionals working on recycling and dredging.

Valorization of Dredged Sediments as Sustainable Construction Resources

Edited by
Amine El Mahdi Safhi

CRC Press
Taylor & Francis Group
Boca Raton London New York

CRC Press is an imprint of the
Taylor & Francis Group, an **informa** business

First edition published 2023
by CRC Press
6000 Broken Sound Parkway NW, Suite 300, Boca Raton, FL 33487-2742

and by CRC Press
4 Park Square, Milton Park, Abingdon, Oxon, OX14 4RN

CRC Press is an imprint of Taylor & Francis Group, LLC

ISBN: 978-1-032-32545-3 (hbk)
ISBN: 978-1-032-32546-0 (pbk)
ISBN: 978-1-003-31555-1 (ebk)

DOI: 10.1201/9781003315551

Typeset in Times
by Deanta Global Publishing Services, Chennai, India

Contents

Preface

The purpose of this book is to provide an overview on the current situation regarding recycling dredged materials (DMs) as a construction material resource, and constitutes a practical guide of valorizing DMs.

Globally, each year, an enormous quantity of DMs is generated for the maintenance of navigable waterways, including harbors, rivers, and reservoirs. In recent years, disposal of DMs on land has increased due to the evolution of regulations which have eliminated disposal in the oceans. The management of those materials has become a serious problem. However, the construction industry became aware of this increase and requested the use of these raw materials to produce concrete and its derivatives. In addition, as a sustainable solution for this issue, several research studies attempted to recycle DMs in the field of civil engineering. Interest in DMs as construction materials has increased as can be seen from the large amount of published literature.

More than 300 research papers were reviewed on the reuse of DMs as a sustainable construction resource and presented in four chapters. The first chapter provides a detailed up-to-date review based on published papers as well as on the industrial chairs and projects on valorizing DMs. The second chapter presents an overview on recycling DMs in non-structural applications, such as road construction, functional soil, and ceramics. The third chapter presents an overview on recycling DMs as filler and aggregates, including fine aggregates and artificial lightweight aggregates. The last chapter presents an overview on the valorization of DMs as a cementitious resource i.e., supplementary cementitious materials, binder for geopolymer, and a raw material for clinker production.

This review revealed that several research studies have been conducted on valorizing DMs in different civil engineering sub-fields. Encouraging findings on the different remediation pathways were reported. The challenges, gap of research, and future scope were presented and discussed. Even though there is a variability of DMs worldwide regarding presence of metallic and metalloid trace elements, and chemical instability, and the fact that in most cases DMs need to be treated before direct use, the end-products are always under the required environmental limits even when DMs are not totally inert. The evaluation of the vast experimental research showed that DMs qualify to be considered as construction material resources. The recycling of these materials will decrease natural resources consumption in the civil engineering field along-side resolving environmental problems.

It is hoped that this book would sustain, to a large extent, the continuous efforts in this field and could be employed to address challenging situations in the management of DMs. Special thanks go to all the contributors for their efforts that led to the production of this distinguished contribution. The authors are grateful to Ms. Sarra Safhi and Pr. Mohsine Mahraj for the proofreading and review, and all CRC Press staff for their extensive efforts throughout this project. It is worth noting that this research received no specific grant from any funding agency in the public, commercial, or not-for-profit sectors. Also, it is acknowledged that no known competing financial interests or personal relationships have influenced the work reported in this paper.

Amine el Mahdi Safhi

Contributors

Mostafa Benzaazoua
Mining Environment and Circular
Economy
Mohammed VI Polytechnic
University
Benguerir, Morocco

Abdelhadi Bouchikhi
CERI- Matériaux et Procédés
Institut Mines Telecom Nord Europe
Douai, Haut-de-France, France

Hassan Ez-Zaki
LCAM - Centre Sciences des
Matériaux
Mohammed V University
Rabat, Morocco

Safaa Mabroum
Mining Environment and Circular
Economy
Mohammed VI Polytechnic
University
Benguerir, Morocco

Patrice Rivard
Centre de Recherche sur
l'infrastructure en béton
Université de Sherbrooke
Sherbrooke, Quebec, Canada

Amine El Mahdi Safhi
Mining Environment and Circular
Economy
Mohammed VI Polytechnic
University
Benguerir, Morocco

Dredged Materials as Construction Resource—Up-to-Date Review

1

Amine El Mahdi Safhi

Contents

DOI: 10.1201/9781003315551-1

1

ABSTRACT

Worldwide, an enormous volume of dredged materials/sediments (DMs) is generated yearly for the maintenance of navigable activities including those involving harbors, rivers, and reservoirs. In recent years, disposal of on-land DMs has increased due to changes in regulations, which have eliminated disposal in the oceans. The management of those materials has become a serious problem. On the other hand, the construction industry has become aware of the increasingly important role of raw materials in producing concrete and its derivatives. To find a sustainable solution to this issue, several research studies have attempted to valorize DMs as construction material. Using the software VOSviewer, a visualization of the interconnections of clusters was conducted on publications related to various valorization pathways of those materials. A literature review on their reuse as a raw materials for civil engineering subfields is presented in this chapter. A detailed database on recycling DMs as construction material is provided. A review on the international and national industrial chairs and projects on the valorization of DMs is conducted, and their achievements are presented and discussed.

1.1 RELEVANCE OF THE TOPIC

The continuous increase of the worldwide population is boosting the necessity to have more construction and buildings, which entails a growing demand on construction materials. A growing demand for non-renewable raw substances has been limited by their scarcity, which in turn has renewed the interest in using recycled materials from several activities (demolition, industrial byproducts, dredged sediments, etc.). Moreover, this practice goes along with sustainability and preservation of the natural environment. On the other hand, dredging operations are mandatory for harbors, as well as reservoirs and rivers, to maintain an acceptable depth for navigation. According to the London Convention, on average 500 million tons (Mt) of permitted dredged material (DMs) are dumped yearly in waters of countries which have signed up to this convention (also known as London Protocol). This protocol bans the dumping of waste with certain specifications at sea to protect the marine environment from human activities. Worldwide, more than one billion cubic meters of DMs were reported to be yearly generated [1], which represents an annual amount of 1300 Mt of DMs, considering a wet density of 1300 kg/m^3. However, in the

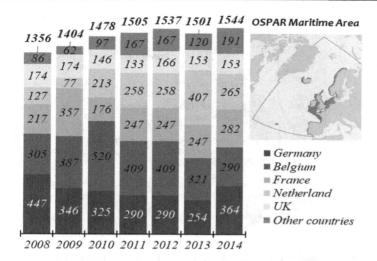

FIGURE 1.1 Total amounts, in Mt, of DMs dumped in the OSPAR Maritime Area per year, per country, over the period 2008–2014 (adapted from the OSPAR Commission website).

OSPAR maritime areas (north-east Atlantic) only, over 1500 Mt of DMs were reported to have been deposited in 2014 (Figure 1.1). Consequently, worldwide DMs are likely to exceed 5000 Mt per year.

These materials are deposed at sea or on land, but oceanic disposal is becoming limited due to changes in the environmental regulations. More and more DMs are now stored on land, which takes up an enormous space. The Waste Framework Directive (2008/98/EC*) provides a hierarchy for prioritizing management of waste streams as presented in Figure 1.2. Minimizing the quantity of generated DMs should be prioritized since prevention is impracticable. DMs should be evaluated for a useful application that will reduce disposal. However, this beneficial use is diverse from one country to another: up to 20% in Ireland, 20–30% in the USA, up to 23% in the Netherlands, about 76% in Spain, and up to 90% in Japan go from land reclamation to engineering uses [2]. In general, the implementation of DMs in remediation activities varies widely depending on the country [3].

Since those materials were considered as an available abundant material, the importance of studying their valorization has increased. Firstly, research focused on the characterization of these materials with a vision to Recycle them as construction materials including using physicochemical, biological,

---

* Directive 2008/98/EC of the European Parliament and of the Council on waste and repealing certain Directives.

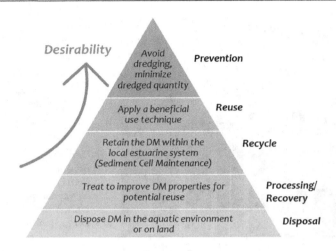

FIGURE 1.2   Hierarchy for prioritizing dredged material management.

and other types of treatments. Other studies focused on the heavy metals and contaminant content [4, 5], effect of calcination on the kinetics of surface area [6], influence of phosphate and thermal treatment on the structure of DMs [7–9], and the rheological and geotechnical characterizations [10, 11]. These studies highlighted the important potential valorization of treated DMs in civil engineering works, which launched an in-depth investigation on their beneficial use as construction materials.

Moreover, the era of digitalizing data and modeling for DMs has arrived. Moghrabi *et al.* (2018) [12] worked on modeling the mechanical strength at 28-d of treated fine DMs. Acceptable accuracies were found despite the poor database (22 points). Chu and Yao (2020) [13] also developed a strength model for concrete based on DMs. In their study, 112 concrete mixtures were investigated, and a strength prediction equation was generated by regression (R^2 of 0.926). Zeraoui *et al.* (2020) [14] created algorithms to optimize the treatment and formulation of DMs for reuse in civil engineering. The program delivers an ideal solution that fits the many technical and environmental criteria of DMs' use while being less expansive. Tran (2021) [15] used an artificial neural network (ANN) for the prediction of unconfined compressive strength (UCS) of stabilized DMs. For this simulation, 52 experimental datasets were collected and employed. More precisely, four variables for input (content of water, cement, air foam, and waste fishing net) created a unique output, which is the compressive strength. The developed ANN model is able to predict the compressive strength with high accuracy when the water content is low. The overall results show that the important input affecting the UCS is in this order:

The content of cement, content of air foam, content of water, and content of waste fishing net. More recently, Wang *et al.* (2022) [16] developed four statistical models aiming to predict the compaction parameters of solidified fine DMs (94 measurements). The variance analysis (ANOVA test) and residual analysis demonstrated that the four models met the statistical criteria and were further validated by using new experimental data with acceptable accuracy.

1.2 STUDIES ON RECYCLING DMS

Using the Scopus website, a database on using dredged materials/sediments as construction material was developed and refined. The refining was done by selecting only publications in English, in subjects related to engineering, environmental science, and materials science, and by selecting the keywords linked to this study. A total of 292 documents were collected and analyzed using the software VOSviewer (v.1.6.18). Based on the collected papers, which had 2382 keywords, only those that had a minimum of five occurrences of a keyword were chosen (188 keywords). A visualization of interconnections of the clusters based on those keywords was carried out as presented in Figure 1.3. The size of the circles correlates to the prevalence of phrases found in published studies in this field.

Four main clusters were formed: Red cluster (56 items), related to cemented and compacted DMs; green cluster (47 items), related to sedimentology and waste management; blue cluster (45 items), related to a recycling in concrete; and yellow cluster (35 items), related to solidification/stabilization (S/S) and the environmental aspect of DMs. To have a better understanding of this, an overlay visualization of the most recent trends in research over time was created (Figure 1.4). It can be noticed that the environmental aspects (contamination, sedimentology, etc.) was investigated at first, followed by the study of their S/S, consolidation, and mechanical behavior. Topics related to the valorization in concrete and cementitious matrix were developed only in very recent years.

It can be noticed that intensive studies were conducted on valorization of DMs as construction materials. In order to collect all of those research studies, a methodology was adopted. First of all, the search was established on looking for keywords from ScienceDirect, Google Scholar, Web of Science, ResearchGate, and Scopus databases. After a careful review of the selected papers, the references used were extracted and added to the database. Then the references used in those papers were reviewed and integrated. This operation was repeated for all the papers until there were no new articles.

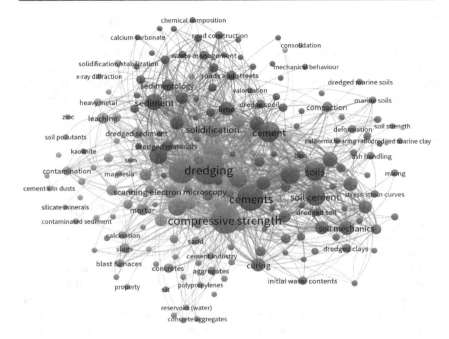

FIGURE 1.3 General network visualization (constructed using VOSviewer): Five clusters, 7180 links, 17906 total link strength.

Publications in French, Spanish, or any other language than English were excluded. This approach makes it possible to collect all the papers about reusing DMs in the field of civil engineering. A total of 306 scientific papers were obtained from 1980 to 2022 and were classified according to the recovery technique and included the title, the author, the country provenance of DMs and their nature, the year of publication, the journal, and the hyperlink (the database will be available under request). All the papers were classified by the recovery pathways. Figure 1.5 summarizes the statistical distribution characteristics of the extracted data.

Figure 1.5 reveals that the interest of research on reusing DMs in a remediation field such as civil engineering has increased during the last three decades. The extracted database using Scopus, which was used for the network visualization, found the same tendency. France is the leading country in this kind of research (30% of publications), followed by China (10% of database), and Algeria with 24 publications (8%). Most of the research studies concerning the DMs' valorization in the civil engineering field were done on road construction techniques (~37%), followed by ceramics and brick production (~21%), and ~14% of the database was about upgrading DMs as supplementary

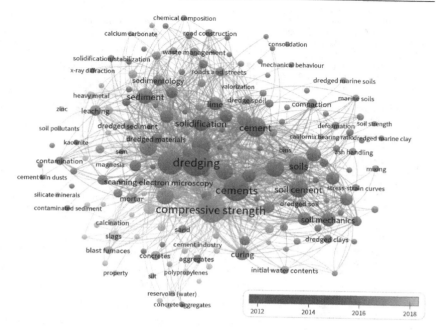

FIGURE 1.4 Overlay visualization (constructed using VOSviewer) depending on publication year.

cementitious materials (SCMs) [17–22]. The majority of the studies was done on marine DMs. However, numerous studies were conducted on reservoir (dams) and fluvial DMs. Only 6% (18 papers) of studies were conducted on DMs from lakes [23–40], and 2% (6 papers) of studies investigated DMs from other provenances e.g., from construction sites [41–43], terrestrial [5], and others [44, 45].

Since the beginning of the 2000s, recycling of DMs as construction material has been studied. At first, researchers tried to stabilize DMs as a functional soil and to use it as backfill or embankments for road bases. Afterwards, extensive studies were made i.e., recycling DMs as raw materials to produce bricks, pavements, and ceramics. Starting from 2010, since the focus was on the physical effects of DMs, several studies focused on valorizing DMs as fine aggregate and even as lightweight aggregate. Giving the fact that those materials are sometimes rich in alumina and silica, depending on their provenance, some studies focused on recycling those materials as geopolymer binder. As researchers became more familiar with the characterization of DMs, in recent years more research has examined their feasibility as SCMs and even as a raw material to produce clinker.

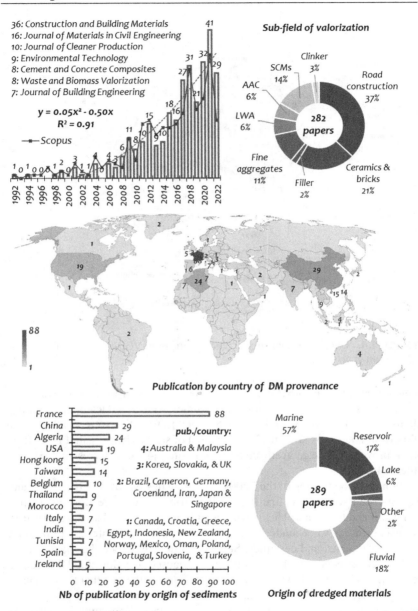

FIGURE 1.5 Statistical distribution of the 306 publications on recycling DMs in the field of civil engineering by country of the DMs' provenance, by year, and field of remediation.

1.3 INDUSTRIAL CHAIRS AND PROJECTS

1.3.1 Introduction

Several industrial chairs and projects have been launched on the feasibility of reusing DMs, promoting their use, or discovering new pathways of recovery. Table 1.1 summarizes the main projects on such valorization. Most of those projects are European, starting mainly in 2007 with *PRISMA, SMOCS, DredgDikes, SETRAMS*, and *CEAMaS*. *PRISMA* aimed to develop new dredging methods, to improve the processing and/or treatment of DMs and dredging, and to find new methods of recycling and reusing in different applications. *SMOCS* focused on the Baltic Sea region, which is known for highly contaminated DMs in ports, estuaries, etc. The project has ten partners from Sweden, Finland, Lithuania, Poland, and Germany, and 20 supporting organizations in these countries as well as Estonia, Denmark, and Russia. *DredgDikes* focused on DMs from the south Baltic region and aimed to recycle those materials for the construction of dikes. The *SETRAMS* project aimed to develop sustainable management practices for marine DMs, considering not only technical parameters but also economic, environmental, social, and regulatory aspects. On the other hand, *CEAMaS* aimed to promote the beneficial reuse of marine DMs in civil engineering applications, in a way that is sustainable, economical, and socially acceptable.

At the end of these projects, more specific ones were launched starting in 2014, including *USAR, SEDIBRIC, VALSE, SURICATE*, and *SEDIASPHALT.* The European project *USAR* focused on the introduction of technologies, methods, and tools for better management and use of DMs by adopting a circular economy approach. To achieve the set targets, it was necessary to identify, analyze, and test potential applications. *SEDIBRIC*, a French project with a much lower budget compared to the other projects, aimed to valorize DMs to produce bricks and tiles. The current projects including *VALSE* and *SURICATE* aim also to valorize those particles and increase their reuse for erosion and flood protection. It is worth mentioning the *Sédimatériaux*, a benchmark approach in the recovery of DMs, as a cooperative approach for the emergence of management and recovery of landfill areas for port and river DMs. This approach aims to improve knowledge and help contractors to innovate building structures based on DMs. This approach has enabled the emergence of new economic channels for the treatment and management of sediments.

A unique chair to date entitled *ECOSED*, standing for the circular ECOnomy of SEDiments, aims to create scientific, technological, and

TABLE 1.1 Overview of the Main Industrial Chairs and Related Information

NO.	PROJECT ACRONYM	PROJECT NAME	PERIOD	TOTAL BUDGET	FINANCING PROGRAM	LEAD PARTNER	COUNTRIES INVOLVED
1	PRISMA	Promoting Integrated Sediment Management	2007–2013	6.41 M€	Interreg IVA 2 Mers Seas Zee ën	Waterwegen en Zeekanal NV	FR|BE|NL|UK
2	SMOCS	Sustainable Management of Contaminated Sediments	2007–2013	3.64 M€	Interreg Baltic Sea Region	Statens Geotekniska Institut	Baltic Sea region
3	DredgDikes	Dredged Materials in Dike Construction	2007–2013	–	EU Interreg IV A	Universität Rostock	South Baltic region
4	SETRAMS	Sustainable Environmental Treatment and Reuse of Marine Sediments	2007–2013	5.01 M€	50% from EU	Association des Ports Locaux de la Manche	FR|UK
5	CEAMaS	Civil Engineering Application for Marine Sediments	2007–2013	4.14 M€	Interreg	CD2E	BE|FR|IE|NL
6	SEDIMEL	Valorization of sediments in the MEL	2013–2020	1.59 M€	EU	MEL (European Metropole of Lille)	FR
7	USAR	Use Sediments as a Resource	2014–2020	4.82 M€	Interreg 2 Mers	Regional Water Authority of Schieland and Krimpenerwaard	FR|BE|NL|UK

(Continued)

TABLE 1.1 (CONTINUED) Overview of the Main Industrial Chairs and Related Information

NO.	PROJECT ACRONYM	PROJECT NAME	PERIOD	TOTAL BUDGET	FINANCING PROGRAM	LEAD PARTNER	COUNTRIES INVOLVED
8	SEDIBRIC	Valorisation de SEDIments en BRIQues et tuiles	2015–2020	972 k€	Interregional State-Regional Plan Contract for the Seine Valley	Grand Port Maritime du Havre	FR
9	SEDIPLAST	Study of a recycled sediment-plastic composite material	2015–2019	1.22 M€	EU	Neo-Eco	FR
10	VALSE	VALorisation of SEdiments	2016–2022	4.17 M€	Interreg France-Wallonie-Vlaanderen	ISSeP	FR|BE
11	SURICATES	Sediment Uses as Resources in Circular and Territorial Economies	2017–2021	5.67 M€	Interreg North-West Europe	University of Lille S&T	FR|IE|NL|UK
12	SEDICIM	SEDiments as a raw material for the production of CEMent	2018–2021	5.20 M€	UE	EQIOM, a CRH Company	FR
13	SEDI-ASPHALT	SEDiments for the production of ASPHALT	2018–2022	5.28 M€	UE	Nord Asphalt	FR
14	NEO'BLOCK	Pavement and blocks based on sediments	2018–2022	2.03 M€	UE	IMT Lille Douai & Neo-Eco	FR

partnership dynamics around the management of DMs with a vision to develop the relevant recovery pathways. The main objective of the *ECOSED Digital 4.0* chair is to optimize the recovery of DMs by measuring, controlling, and configuring the characteristics of materials and industrial processes. Based on field demonstrators produced in partnership with the chair's partners, the work aims to allow the operational optimization of DMs' recovery channels. This chair gave birth to several projects such as *SEDIROUTE, SEDIMEL* (use of fluvial DMs for road construction), *SEDICIM* (feasibility of DMs for cement production), *NEO'BLOCK*, and *MODELISED* (MODELing and study of industrial processes applied to SEDiment management).

1.3.2 Outcomes and Achievements

The general outcome of the above-mentioned projects led to diverse determinations including reports and studies, new models and techniques, pilot studies, and testing. The reports and studies gave an overview of all techniques in the field of dredging and dewatering of DMs regarding environment, quality and quantity, costs and benefits, and applicability in various areas. The description of new treatment methods and techniques was provided for each project. Some even provided an evaluation of the effects and benefits of the reuse of DMs in habitat creation, dikes and banks, roads, and site preparation of land. More specifically, the *DredgDikes* project designed the South Baltic guidelines for the use of DMs, coal combustion products, and geosynthetics in dike building [46]. The *Sédimatériaux* approach produced three methodological guides on road technique, landscaping, and concrete. They constitute a real 'manual' for all operators confronted with the problems of land management of port and river DMs.

New models and procedures, including models of decision-making, were developed and tested at several places for the prospects of the treatment and reuse of DMs. In addition, adopted techniques and installations for the dewatering of DMs described and assessed new approaches for the reuse of DMs in island construction for the establishment of new habitats. For example, the *SMOCS* project has developed a toolbox for different decision situations that comprise the assessment methods (Life Cycle Analysis and Risk Assessment) and decision support tools (Multi Criteria Decision Analysis). Also, one of the outcomes of *USAR* is a software for the optimization of concrete, dike, road, etc. [47]. The software contains a database of DMs from partner countries with geolocation data for the dredging and storage areas. Using this data, the software enables the calculation of transport distance and treatment processes of DMs according to the regulations of those countries. A mix design law was elaborated to predict the different physical and chemical treatments required for each type of application area. The results proved that the solution proposed by the software was effective during the construction of a road in the city of Lille (France). The *CEAMaS* project delivered a practical toolbox that contains a multi-criteria decision-making tool, EU evaluation of legislation status for sediments tools, a database of their characteristics, and online GIS (WebGIS) that allows exploring and comparing topographic data of Europe on a map.

Pilot testing is very important to concretize the recovery of DMs. Several projects were carried out as part of the *Sédimatériaux* approach, such as:

- *Blocs of concrete* (2009–2013): A project of the Grand Port Maritime of Dunkirk (GPMD) (150 k€), where 110 DMs-based concrete blocks were cast to reinforce the dikes. The DMs were treated, and a binder based on sulfoaluminous calcium clinker was incorporated. On average, the concrete blocks integrated between 12% to 20% of DMs, which corresponds to approximately 0.5 m^3 of DMs per m^3 of concrete.
- *Landscapeeco-model* (2009–2013): With a budget of 3.17 M€, this project, also of GPMD, consists of the creation of a landscaped eco-model whose core is made up of DMs. The mound, 500 meters long and varying in height from five to seven meters, is covered with a 25 cm layer of topsoil to ensure the stability of the DMs.
- *Freycinet 12 road* (2009–2013): Another project of GPMD, with a budget of 275 k€, a road was constructed by COLAS Nord Picardie. This application used 1 m^3 of DMs for 10 m^2 of road. Thus, 450 m^3 of dry DMs, representing 1800 m^3, were used.
- *VNF* (2013–2019): In this 1.29 M€ project, two DMs-based concretes were tested on a field scale: The manufacture of prefabricated concrete gabion mattresses and the manufacture of ready-to-use concrete coping beams. All this ensures good stability of the bank.
- *SEDIMEL* (2013–2020): Three test sites in the Lille metropolitan area used 500 tons of DMs to include the creation of a parking and a buffer basin (Tourcoing), creation of a buffer basin in hollow concrete under the bus station and parking (Leers), and self-consolidating DMs-based grout for trench backfill (Bondues).
- *Barbieux park* (2016–2018): Project by the city of Roubaix financed by the Region funds (846 k€). In this project, 10000 m^3 of DMs were recovered in a short circuit. Several pilot studies were made with different formulations. After one year of monitoring, the adopted mix design comprises 90% DMs and 10% bentonite.
- *VALODIGUE* (2017–2020): Project of *CAPSO*, with a budget of 292 k€ financed up to 50% by the region Hauts-de-France, aimed to recover DMs in flood control structures (dikes) passing by i) a

detailed characterization; ii) mix design of a dike incorporating DMs; and iii) evaluation of the environmental impact of the final product.

- *OPTISED* (2017–2020): Project of 300 k€, financed by the water agency of Artosi–Picardie region, for reusing the fluvial DMs (Nord-Pas-de-Calais, France) and industrial waste (waste glass, fly ash, silica fumes, etc.) in a cementitious matrix including blended binders, sand substitution, and alkali-activation binder.
- *NEO'BLOCK* (2018–2022): Pavement based on 50% of DMs as binder of fine aggregate: three experimental pilot studies, each made up of nearly 200 interlocking pavers, will be conducted to assess their behavior in-situ in a real functional environment.
- *SEDENOV* (2021–2023): 700 k€ project for valorization of DMs from Rouen region in road construction. The project is conducted by *Neo-Eco*, IMT Nord-Europe, and Rouen region.

1.3.3 Overview

Alongside with the research studies, the industrial chairs are important, and the projects concretize the recovery and the conducted studies. Road construction, building artificial parks, and parking are wide applications in civil engineering and infrastructure development. Other non-structural application pathways were applied including brick and dike production. However, the realized projects do not follow on from the studies conducted in the research centers, i.e., no pilot projects were conducted on the valorization in structural applications such as cement and clinker production. More efforts must be done in such recovery.

1.4 PREVIOUS REVIEW PAPERS

Seven review papers were published on using DMs for construction purposes. Burnett and Whiteside (1992) [48] investigated the suitability of using dredged sand and gravel from Hong Kong harbors and developed a scheme assessing the reuse of the deposits according to several standards. The addressed scheme includes knowing the origin of DMs, their class (sand/gravel), their mass in term of cementation (indurated, non-/well cemented) and compaction, their quality in term of color, grading, constitution, composition, texture, and the shape of particles. The limited tests used indicated that the alluvial sand

and gravel from the Tsing Yi location might work well as fine aggregate in structural concrete, general fill, bituminous road pavement, and for masonry grouts and mortars. Yozzo et al. (2004) [49] reviewed the Dredged Material Management Plan that was conducted by the US Army Corps of Engineers (USACE). Yearly, more than 24 Mm³ (million-meter-cubed) DMs from the port of New York/New Jersey could be placed for creation of artificial reefs and shoals, oyster reef restoration, bathymetric recontouring, augmentation of intertidal marshes and mud flats, filling up dead-end canals and basins, and development of wildlife islands—all of which are examples of restoration projects. Ultimately, the estimated placement cost should be balanced by the environmental benefits as expected. Parson and Swafford (2012) [50] reported the beneficial reuse of DMs from the Gulf of Mexico, which was also done by USACE. For a yearly dredged quantity of 76 Mm³, early coordination and frequent cooperation of interdisciplinary interagency teams was highly recommended, alongside organizing annual dredging conferences to lay out timetables and identify potential sources of DMs. Sheehan and Harrington (2012) [2] examined the key components of the Irish dredging sector in relation to international practice and standards, and noted that global knowledge of the beneficial uses of DMs had been raised. The governing legislation was analyzed and critically appraised. Dauji (2017) [51] presented a review of approaches employed to recycle DMs in the construction industry and biodiesel production. He reported that the main consideration for the reuse of DMs includes grain size distribution, mechanical properties, durability concerns, environmental issues, economic considerations, and thermal expansion of the DMs. Ean et al. (2018) [52] conducted a review on the characterization of DMs for the production of green bricks. They concluded that the method of producing sediment bricks (cementing or other) depends on the various characteristics of DMs. They recommended conducting leaching tests for the end-product to guarantee that the leached heavy metals do not exceed the legal limitations. In her review on beneficial use of DMs in geotechnical engineering, Balkaya (2018) [53] encourages such valorization when the physiochemical properties are good and the contamination levels of the DMS are within acceptable limits. Junakova and Junak (2019) [54] reviewed the potential paths of the management of DMs. The focus was made on the possibility of reusing DMs as a raw material for construction. The findings of selected research that focused on DMs' long-term reuse as raw materials in the production of lightweight aggregates, bricks and concrete were reported. The results revealed that such applications are viable. Nevertheless, there are certain economic, technological, and social constraints. Another technical review was done by Amar et al. (2020) [55] on the reuse of DMs as SCMs. The impact of sediment contaminants in cement matrixes, different treatment processes, quantification of chemical activity, and the durability and environmental impact were reported. The study proposed to qualify the treated DMs as SCMs using the performance approach according to the NF EN 206-1/CN standard [56]. A very recent review paper was published by Zhang *et al.* (2021) [57] on the sustainable remediation of contaminated DMs. A detailed up-to-date review of the contaminants in different DMs was presented, alongside the immobilization and removal mechanisms used for their treatment. Some examples were given for resource utilization of DMs as construction materials, alongside an introduction to evaluation methods for different ex-situ DMs' remediation technologies being addressed.

1.5 CONCLUSION: SO, WHAT IS THE NEED FOR ANOTHER REVIEW?

Clearly, there are few literature reviews compared to the published research that suggest the conduction of more detailed up-to-date reviews (Figure 1.6). Furthermore, the previous review papers either focus on regional case studies, limited either to general beneficial use of DMs or to particular applications (SCMs), or all of those mentioned above. A review that presents a detailed critical review of recycling DMs as a building material resource in different

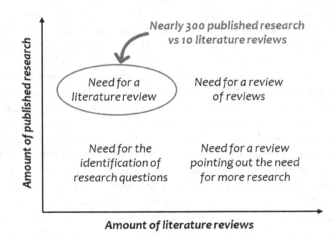

FIGURE 1.6 A conceptual diagram depicting the necessity for various sorts of literature reviews based on the number of published research articles and literature reviews, adapted from the open access paper of Pautasso M., (2013), PLoS Comput Biol 9(7): e1003149. Copyright (2013). Figure 1. A. [58].

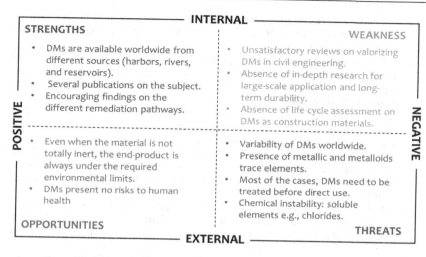

FIGURE 1.7   SWOT analysis on DMs as sustainable construction material.

fields of civil engineering is needed—a review that includes recycling of DMs as functional soil, backfill or embankments for road bases, for production of bricks, pavements, and ceramics, as fine and lightweight aggregate, geopolymer binder, SCMs, and even as a raw material to produce clinker. A SWOT (Strengths, Weaknesses, Opportunities, and Threats) analysis to identify research gaps is given in Figure 1.7, considering the growing interest in valorizing DMs in the civil engineering field, due to sustainability perspectives. The next chapters aim to (i) highlight the evolution of the use of DMs as a construction material, (ii) support and guide managers and decision makers during the development of their valorization projects, (iii) propose a harmonized methodological and scientific framework for the recovery of DMs on land, (iv) identify and develop recovery channels adapted to any type of DMs, and (v) produce the scientific data necessary for the development regulatory framework at the national level.

REFERENCES

1. M. Benzerzour, M. Amar, N.-E. Abriak, New experimental approach of the reuse of dredged sediments in a cement matrix by physical and heat treatment, *Constr. Build. Mater.* 140 (2017) 432–444. https://doi.org/10.1016/j.conbuildmat.2017.02.142.

2. C. Sheehan, J. Harrington, Management of dredge material in the Republic of Ireland: A review, *Waste Manag.* 32 (2012) 1031–1044. https://doi.org/10.1016/j.wasman.2011.11.014.

3. A.P. Lehoux, K. Petersen, M.T. Leppänen, I. Snowball, M. Olsen, Status of contaminated marine sediments in four Nordic countries: Assessments, regulations, and remediation approaches, *J. Soils Sediments.* 20 (2020) 2619–2629. https://doi.org/10.1007/s11368-020-02594-3.

4. F. MacFarlane, E. Henry, C. Boulemia, Management approach of contaminated dredged materials for valorisation in civil engineering applications to dunkirk seaport, in: Dundee, UK, 2003: pp. 143–154.

5. X. Capilla, C. Schwartz, J.-P. Bedell, T. Sterckeman, Y. Perrodin, J.-L. Morel, Physicochemical and biological characterisation of different dredged sediment deposit sites in France, *Environ. Pollut.* 143 (2006) 106–116. https://doi.org/10.1016/j.envpol.2005.11.007.

6. J. Ramaroson, J.-L. Dirion, A. Nzihou, G. Depelsenaire, Characterization and kinetics of surface area reduction during the calcination of dredged sediments, *Powder Technol.* 190 (2009) 59–64. https://doi.org/10.1016/j.powtec.2008.04.094.

7. S. Kribi, J. Ramaroson, A. Nzihou, P. Sharrock, G. Depelsenaire, Laboratory scale study of an industrial phosphate and thermal treatment for polluted dredged sediments, *Int. J. Sediment Res.* 27 (2012) 538–546. https://doi.org/10.1016/S1001-6279(13)60011-6.

8. J. Ramaroson, M. Dia, J.-L. Dirion, A. Nzihou, G. Depelsenaire, Thermal treatment of dredged sediment in a rotary kiln: Investigation of structural changes, *Ind. Eng. Chem. Res.* 51 (2012) 7146–7152. https://doi.org/10.1021/ie203023k.

9. M. Dia, R. Zentar, N. Abriak, A. Nzihou, G. Depelsenaire, A. Germeau, Effect of phosphatation and calcination on the environmental behaviour of sediments, *Int. J. Sediment Res.* 34 (2019) 486–495. https://doi.org/10.1016/j.ijsrc.2018.10.002.

10. I. Bel hadj ali, Z. Lafhaj, M. Bouassida, I. Said, Characterization of Tunisian marine sediments in Rades and Gabes harbors, *Int. J. Sediment Res.* 29 (2014) 391–401. https://doi.org/10.1016/S1001-6279(14)60053-6.

11. H. Wang, R. Zentar, D. Wang, Rheological characterization of fine-grained sediments under steady and dynamic conditions, *Int. J. Geomech.* 22 (2022) 04021260. https://doi.org/10.1061/(ASCE)GM.1943-5622.0002243.

12. I. Moghrabi, H. Ranaivomanana, F. Bendahmane, O. Amiri, D. Levacher, Modelling the mechanical strength development of treated fine sediments: A statistical approach, *Environ. Technol.* 40 (2019) 1890–1909. https://doi.org/10.1080/09593330.2018.1432697.

13. S.H. Chu, J.J. Yao, A strength model for concrete made with marine dredged sediment, *J. Clean. Prod.* 274 (2020) 122673. https://doi.org/10.1016/j.jclepro.2020.122673.

14. M. Zeraoui, W. Benzerzour, R.M. Maherzi, N.-E. Abriak, New software for the optimization of the formulation and the treatment of dredged sediments for utilization in civil engineering, *J. Soils Sediments.* 20 (2020) 2709–2716. https://doi.org/10.1007/s11368-020-02605-3.

15. V.Q. Tran, Compressive strength prediction of stabilized dredged sediments using artificial neural network, *Adv. Civ. Eng.* 2021 (2021) e6656084. https://doi.org/10.1155/2021/6656084.

16. H. Wang, R. Zentar, D. Wang, Predicting the compaction parameters of solidified dredged fine sediments with statistical approach, *Mar. Georesources Geotechnol.* 0 (2022) 1–16. https://doi.org/10.1080/1064119X.2021.2023827.

17. el M. Safhi, P. Rivard, A. Yahia, M. Benzerzour, K.H. Khayat, Valorization of dredged sediments in self-consolidating concrete: Fresh, hardened, and microstructural properties, *J. Clean. Prod.* 263 (2020) 121472. https://doi.org/10.1016/j.jclepro.2020.121472.

18. el M. Safhi, M. Benzerzour, P. Rivard, N.-E. Abriak, I. Ennahal, Development of self-compacting mortars based on treated marine sediments, *J. Build. Eng.* 22 (2019) 252–261. https://doi.org/10.1016/j.jobe.2018.12.024.

19. el M. Safhi, M. Benzerzour, P. Rivard, N.-E. Abriak, Feasibility of using marine sediments in SCC pastes as supplementary cementitious materials, *Powder Technol.* (2018). https://doi.org/10.1016/j.powtec.2018.12.060.

20. el M. Safhi, P. Rivard, A. Yahia, K. Henri Khayat, N.-E. Abriak, Durability and transport properties of SCC incorporating dredged sediments, *Constr. Build. Mater.* 288 (2021) 123116. https://doi.org/10.1016/j.conbuildmat.2021.123116.

21. M. Amar, M. Benzerzour, A.E.M. Safhi, N.-E. Abriak, Durability of a cementitious matrix based on treated sediments, *Case Stud. Constr. Mater.* (2018). https://doi.org/10.1016/j.cscm.2018.01.007.

22. A. Bouchikhi, A. el M. Safhi, P. Rivard, R. Snellings, N.-E. Abriak, Fluvial sediments as SCMs: Characterization, pozzolanic performance, and optimization of equivalent binder, *J. Mater. Civ. Eng.* 34 (2022) 04021430. https://doi.org/10.1061/(ASCE)MT.1943-5533.0004071.

23. C.L. Zhang, Q.S. Liu, J.B. Liu, Field Plate Load tests on dredged sediment dump pond with cement solidified crust above, *Adv. Mater. Res.* 255–260 (2011) 2751–2755. https://doi.org/10.4028/www.scientific.net/AMR.255-260.2751.

24. X. He, Y. Chen, Y. Wan, L. Liu, Q. Xue, Effect of curing stress on compression behavior of cement-treated dredged sediment, *Int. J. Geomech.* 20 (2020) 04020204. https://doi.org/10.1061/(ASCE)GM.1943-5622.0001857.

25. T. Chompoorat, T. Thepumong, S. Taesinlapachai, S. Likitlersuang, Repurposing of stabilised dredged lakebed sediment in road base construction, *J. Soils Sediments.* 21 (2021) 2719–2730. https://doi.org/10.1007/s11368-021-02974-3.

26. S. Mohammad, W. Akram, S.A. Mirza, Geotechnical characterization of dredged material and effect of lime stabilisation on its strength characteristics, *Appl. Mech. Mater.* 877 (2018) 289–293. https://doi.org/10.4028/www.scientific.net/AMM.877.289.

27. T. Chompoorat, S. Likitlersuang, T. Thepumong, W. Tanapalungkorn, P. Jamsawang, P. Jongpradist, Solidification of sediments deposited in reservoirs with cement and fly ash for road construction, *Int. J. Geosynth. Ground Eng.* 7 (2021) 85. https://doi.org/10.1007/s40891-021-00328-0.

28. T. Chompoorat, K. Thanawong, S. Likitlersuang, Swell-shrink behaviour of cement with fly ash-stabilised lakebed sediment, *Bull. Eng. Geol. Environ.* 80 (2021) 2617–2628. https://doi.org/10.1007/s10064-020-02069-2.

29. J. Wu, G. Leng, X. Xu, Y. Zhang, X. Lao, K. Li, Preparation and properties of ceramic facing brick from East-lake sediment, *J. Wuhan Univ. Technol.-Mater Sci Ed.* 27 (2012) 154–159. https://doi.org/10.1007/s11595-012-0427-1.

30. Y.M. Zhang, L.T. Jia, H. Mei, Q. Cui, P.G. Zhang, Z.M. Sun, Fabrication, microstructure and properties of bricks fired from lake sediment, cinder and sewage sludge, *Constr. Build. Mater.* 121 (2016) 154–160. https://doi.org/10.1016/j.conbuildmat.2016.05.155.

31. Y. Rani, D. Dheeravath, V. Reddy, S. Poodari, H. Vurimindi, Central bringing excellence in open access utilization of polluted dredged sediment for making of bricks, *JSM Chem.* 5 (2017).

32. X. Peng, Y. Zhou, R. Jia, W. Wang, Y. Wu, Preparation of non-sintered lightweight aggregates from dredged sediments and modification of their properties, *Constr. Build. Mater.* 132 (2017) 9–20. https://doi.org/10.1016/j.conbuildmat.2016.11.088.

33. Y. Peng, X. Peng, M. Yang, H. Shi, W. Wang, X. Tang, Y. Wu, The performances of the baking-free bricks of non-sintered wrap-shell lightweight aggregates from dredged sediments, *Constr. Build. Mater.* 238 (2020) 117587. https://doi.org/10.1016/j.conbuildmat.2019.117587.

34. Q. Wan, C. Ju, H. Han, M. Yang, Q. Li, X. Peng, Y. Wu, An extrusion granulation process without sintering for the preparation of aggregates from wet dredged sediment, *Powder Technol.* 396 (2022) 27–35. https://doi.org/10.1016/j.powtec.2021.10.030.

35. R. Kou, M.-Z. Guo, L. Han, J.-S. Li, B. Li, H. Chu, L. Jiang, L. Wang, W. Jin, C. Sun Poon, Recycling sediment, calcium carbide slag and ground granulated blast-furnace slag into novel and sustainable cementitious binder for production of eco-friendly mortar, *Constr. Build. Mater.* 305 (2021) 124772. https://doi.org/10.1016/j.conbuildmat.2021.124772.

36. X. Yang, L. Zhao, M.A. Haque, B. Chen, Z. Ren, X. Cao, Z. Shen, Sustainable conversion of contaminated dredged river sediment into eco-friendly foamed concrete, *J. Clean. Prod.* 252 (2020) 119799. https://doi.org/10.1016/j.jclepro.2019.119799.

37. H. Li, F. Huang, Y. Xie, Z. Yi, Z. Wang, Effect of water–powder ratio on shear thickening response of SCC, *Constr. Build. Mater.* 131 (2017) 585–591. https://doi.org/10.1016/j.conbuildmat.2016.11.061.

38. T.T.M. Nguyen, S. Rabbanifar, N.A. Brake, Q. Qian, K. Kibodeaux, H.E. Crochet, S. Oruji, R. Whitt, J. Farrow, B. Belaire, P. Bernazzani, M. Jao, Stabilization of silty clayey dredged material, *J. Mater. Civ. Eng.* 30 (2018) 04018199. https://doi.org/10.1061/(ASCE)MT.1943-5533.0002391.

39. K.H. Khayat, J. Assaad, J. Daczko, Comparison of field-oriented test methods to assess dynamic stability of self-consolidating concrete, *ACI Mater. J.* 101 (2004). https://doi.org/10.14359/13066.

40. D. Wang, J. Xiao, X. Gao, Strength gain and microstructure of carbonated reactive MgO-fly ash solidified sludge from East Lake, China, *Eng. Geol.* 251 (2019) 37–47. https://doi.org/10.1016/j.enggeo.2019.02.012.

41. W.Z. Zhou, H. Wei, T. Liu, D. Zou, H. Guo, A novel approach for recycling engineering sediment waste as sustainable supplementary cementitious materials, *Resour. Conserv. Recycl.* 167 (2021) 105435. https://doi.org/10.1016/j.resconrec.2021.105435.

42. Y.T. Kim, J. Ahn, W.J. Han, M.A. Gabr, Experimental evaluation of strength characteristics of stabilized dredged soil, *J. Mater. Civ. Eng.* 22 (2010) 539–544. https://doi.org/10.1061/(ASCE)MT.1943-5533.0000052.

43. L. Zhang, X. Jia, G. Yang, D. Wang, High-volume upcycling of waste sediment and enhancement mechanisms in blended grouting material, *Adv. Mater. Sci. Eng.* 2022 (2022) e3225574. https://doi.org/10.1155/2022/3225574.

44. S. Bhairappanavar, R. Liu, A. Shakoor, Eco-friendly dredged material-cement bricks, *Constr. Build. Mater.* 271 (2021) 121524. https://doi.org/10.1016/j.conbuildmat.2020.121524.

45. M.P.D. Ingunza, K.L. de A. Pereira, O.F. dos Santos Junior, Use of sludge ash as a stabilizing additive in soil-cement mixtures for use in road pavements, *J. Mater. Civ. Eng.* 27 (2015) 06014027. https://doi.org/10.1061/(ASCE) MT.1943-5533.0001168.

46. F. Saathoff, S. Cantré, Z. Sikora, Universität Rostock, eds., *South Baltic Guideline for the Application of Dredged Materials, Coal Combustion Products and Geosynthetics in Dike Construction: [Project] Dredg Dikes*, Universität Rostock Agrar- und Umweltwissenschaftliche Fakultät [u.a.], Rostock, 2015.

47. M. Zeraoui, W. Benzerzour, R.M. Maherzi, N.-E. Abriak, New software for the optimization of the formulation and the treatment of dredged sediments for utilization in civil engineering, *J. Soils Sediments.* (2020). https://doi.org/10.1007/ s11368-020-02605-3.

48. A.D. Burnett, P.G.D. Whiteside, Dredged sand and gravel for construction purposes: An assessment procedure and Hong Kong case study, *J. Coast. Res.* 8 (1992) 105–124.

49. D.J. Yozzo, P. Wilber, R.J. Will, Beneficial use of dredged material for habitat creation, enhancement, and restoration in New York–New Jersey Harbor, *J. Environ. Manage.* 73 (2004) 39–52. https://doi.org/10.1016/j.jenvman.2004.05 .008.

50. L.E. Parson, R. Swafford, Beneficial use of sediments from dredging activities in the Gulf of Mexico, *J. Coast. Res.* 60 (2012) 45–50. https://doi.org/10.2112/SI _60_5.

51. S. Dauji, Disposal of sediments for sustainability: A review, *Int. J. Econ. Energy Environ.* 2 (2017) 96. https://doi.org/10.11648/j.ijeee.20170206.12.

52. L.W. Ean, M.A. Malek, B.S. Mohammed, C.-W. Tang, P.X.H. Bong, A review on characterization of sediments for green bricks production, *Int. J. Eng. Technol.* 7 (2018) 41–47. https://doi.org/10.14419/ijet.v7i4.35.22319.

53. M. Balkaya, Beneficial use of dredged materials in geotechnical engineering, in: N. Balkaya, S. Guneysu (Eds.), *Recycling and Reuse Approaches for Better Sustainability*, Springer International Publishing, Cham, 2019: pp. 21–38. https:// doi.org/10.1007/978-3-319-95888-0_3.

54. N. Junakova, J. Junak, Alternative reuse of bottom sediments in construction materials: Overview, *IOP Conf. Ser.: Mater. Sci. Eng.* 549 (2019) 012038. https:// doi.org/10.1088/1757-899X/549/1/012038.

55. M. Amar, M. Benzerzour, J. Kleib, N.-E. Abriak, From dredged sediment to supplementary cementitious material: Characterization, treatment, and reuse, *Int. J. Sediment Res.* 36 (2021) 92–109. https://doi.org/10.1016/j.ijsrc.2020.06.002.

56. AFNOR, NF EN 206/CN - Concrete - Specification, performance, production and conformity - National addition to the standard NF EN 206 (2014).

57. Y. Zhang, C. Labianca, L. Chen, S. De Gisi, M. Notarnicola, B. Guo, J. Sun, S. Ding, L. Wang, Sustainable ex-situ remediation of contaminated sediment: A review, *Environ. Pollut.* 287 (2021) 117333. https://doi.org/10.1016/j.envpol.2021.117333.

58. M. Pautasso, Ten simple rules for writing a literature review, *PLOS Comput. Biol.* 9 (2013) e1003149. https://doi.org/10.1371/journal.pcbi.1003149.

Recycling Dredged Materials for Road Construction and Ceramic Production

2

Amine El Mahdi Safhi, Safaa Mabroum,
Abdelhadi Bouchikhi, Hassan Ez-Zaki,
Mostafa Benzaazoua

Contents

DOI: 10.1201/9781003315551-2

23

ABSTRACT

Consumption of non-renewable natural materials is highest in the field of civil engineering, being used for road construction. France needs ~400 million tons (Mt) to build roads each year and ~15 Mt for the maintenance of road networks. The use of alternative materials such as non-submerged marine and fluvial dredged materials (DMs) mainly for earthworks and the earth's sub-layer is increasing. In general, socio-economic technical requirements are the key parameters to be controlled. In some projects, DMs were used as level corrector or reinforcement spoil in the case of bridges and digs. Prior to any methodological approach of valorization in road construction, a geotechnical characterization is very important. It is this first step that will help to recognize potential treatments depending on the properties of the DMs. This occurs in ceramic production as well; the variation in the physio-chemical properties created the need of adding other wastes or raw materials. Overall, the results are encouraging and will positively affect the economic and environmental aspects.

2.1 VALORIZATION OF DMS AS GEOMATERIALS FOR ROAD CONSTRUCTION

2.1.1 Structure of a Roadway

Roadway engineering consumes a huge quantity of non-renewable natural materials. Thus, a wide range of material is needed to fill the gap in this field. A good solution to this issue is the use of alternative materials such as the non-submerged dredged materials (DMs). Indeed, stabilizing DMs for road construction or functional soils was one of the earliest remediation pathways. For a feasible recovery of local alternative materials, in particular DMs, socio-economic technical requirements are the key parameters to be controlled. From a technical point of view, the majority of roads are made from the superposition of layers with distinct mechanical characteristics. In a pavement structure, there are four levels of the main layers: (i) The compacted subgrade, (ii) the sub-base course (sub-formation), (iii) the base layer (capping layer), and (iv) the surface layers (layers of bearing and binding). The majority of studies conducted on recycling DMs as a road construction were rather focused on valorization in sub-layers, which requires a relatively lower performance. These layers and the demand for higher volume of raw materials are important compared to the other layers, i.e., capping layer, sub-base layer, and base layer. Recently, more research was conducted on DMs as a compound in the surface course part, in particular to substitute the fine part as well as to replace the sandy fraction. The quantity of DMs valued in this part remains negligible compared to the other layers mentioned above (Figure 2.1).

2.1.2 Characterization of the Used Sediments

Based on their nature, DMs are classified in most cases as fine materials and sometimes rich in organic matter (OM). Treatment with hydraulic binders (cement or lime) or other material is necessary to solidify the matrix. Previously, to avoid the harmful effect of the presence of DMs (which can be of acid nature), lime treatment ensures a basic pH in the mixture where the hydraulic binders can react easily. The presence of OM is the main issue that tends to decrease the performances. The presence of salt also decreases the performance of the structure, particularly the sulphate which favors the formation of thaumasite, responsible for swelling and degradation of the structures. Another critical parameter is obtaining the granularity, i.e., an adaptation of the particle size by a granular correction could be essential. The use of materials in road engineering must be based on criteria following adaptive guidelines or traffic load. For this, geotechnical parameters of materials are key elements in their classification. Depending on the mineralogical compositions and the OM content, the geotechnical classifications of materials change. The environmental behavior is also a critical parameter that influences the acceptability of DMs in road construction (Figure 2.2).

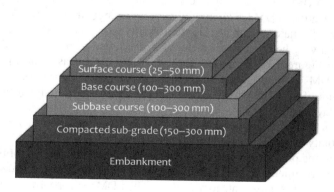

FIGURE 2.1   Standard section of a road structure.

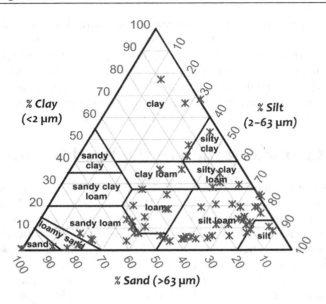

FIGURE 2.2 Classification of 74 DMs in sand-silt-clay triangular diagram from 44 papers [1–44].

Among the 102 papers published on recycling DMs as a construction material for road application, Table 2.1 presents the main reported physical properties. The water content in DMs was found to be varying in a wide range of 5–235% with an average density of 2.57 g/cm^3 and an average particle size of 76% under 80 μm. The OM content was found in a range of 1.7–16% and a plasticity index of 7.7–80 with an average of 30. The average value of methylene blue (VBS) was ~2.23 which corresponds to a loamy clay soil, however, the classification of DMs (#74) in sand-silt-clay triangular diagram (Figure 2.3) shows that the majority can be classified as silty loam materials. Nevertheless, DMs are a very heterogeneous material, some found to be sandy and clayey materials. Figure 2.3 corroborates those findings; in the Casagrande plasticity diagram DMs were aligned with so called A-line. The majority of DMs have liquid limit of 40–60% which corresponds to a medium to high plasticity and a silty–clayey nature.

When DMs are rich in clay and OM, the water retention increases, and the plasticity is affected by the presence of swelling clays and OM [73, 74]. Moreover, the presence of pollutants in DMs affects the reactivity of the used binder for treatment e.g., the effects of heavy metal ions on the hydration of cement [75, 76]. A large category of the studied DMs is concerned with these factors including OM, swelling clays, heavy metals, etc. For this, research

TABLE 2.1 Properties of DMs Used in Road Techniques

	W, %	OM AT 450 °C, %	Ps, g/cm³	VBS	<80 µm, %	IP
Nb of data	85	56	75	39	30	98
Min	5.20	1.72	2.13	0.10	3.00	7.70
Max	235	16.1	3.02	4.41	99.6	80.0
Average	87.1	6.99	2.57	2.23	76.4	30.4
S.D.	56.6	4.10	0.16	1.13	25.4	17.0

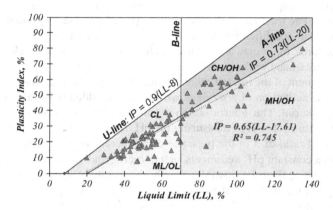

FIGURE 2.3 Plasticity diagram of Casagrande of 98 DMs from 63 papers [1–7, 9–12, 14, 17–20, 22–27, 30–38, 40–42, 44, 45, 45–72]. Legend: C: Clay, M: Silt, O: Organic, H: High plasticity, L: Low plasticity.

works focused on the behavior of DMs in the road-base construction. Several treatments were investigated depending on the chemical and mineralogical properties, the content of pollutants, and the OM.

2.1.3 Elaboration of the Road Material

The mixture design study aims to examine the feasibility of stabilizing DMs or replacing a fraction of sand used in road materials. Due to its availability, the use of dredged sand presents an attractive economic solution for improving the mechanical characteristics of stabilized soils. The granular correction is generally required, and the granular correctors choice must satisfy the economical, technical, and environmental constraints. Dubois *et al.* (2009–2011) [4, 77] have proposed a methodology for the choice and addition of the granular corrector mainly based on: i) Limitation of the proportion of the fine fraction in the mixtures, ii) limiting the amount of OM, iii) optimization of the amount of fine DMs, and iv) optimization of particle size distribution (PSD). Optimizing the PSD is a very important step in building the granular skeleton of road materials.

The mixture of sediments and dredged sand must then be treated with quicklime and/or hydraulic binder. Hydraulic binder is a necessary constituent in the processing of materials for their reuse in road engineering. Its ability to agglomerate aggregates gives the material a permanent cohesion, which depends on the nature of treated materials, the type of binder, the dosage, the achieved compactness, and the temperature of the early and later treatment stages. The treatment of DMs using a hydraulic binder is particularly for upgrading their solidification. Quicklime is commonly employed in stabilization procedures, but it may also be used in hydraulic binders as an addition. As part of the treatment of fine sediments, the benefit of lime consists in reducing the water content of the treated products and also in producing lime hydroxide, allowing the activation of pozzolans which may be added if not already present in the material. The determination of lime content is governed by the limit fixation test that consists of measuring the pH of the sediment' suspensions, at a liquid/solid ratio of 5 in the presence of an increasing percentage of quicklime up to a constant pH. Sediments rich in racing and vegetation require special treatment against the possible growth of vegetation; stability over time is necessary before finalizing the construction of the road. Figure 2.4 illustrates this approach for the valorization of DMs in road construction, inspired from Maherzi (2010) [78].

2.1.4 Conclusions

The utilization of DMs in soil engineering has increased and shows possibilities for the utilization of sediments in stabilized soil blocks. From the results and discussion, experimental detailed characterization of DMs is necessary before considering any treatment. The higher water content of these materials makes their dehydration necessary. The overall results showed a wide variability of DMs i.e., fine to sandy material with low to high OM and clay activity. These properties directly affect the geotechnical properties of DMs and thus those of the functional soils. Hence, more investigation on the durability of the soil blocks is required to increase the application of stabilized blocks in the field. More study is equally needed to determine the appropriateness of various DMs for treatment as well as the conditions under which the addition of binder may result the treated DMs acceptable for subgrade layer application.

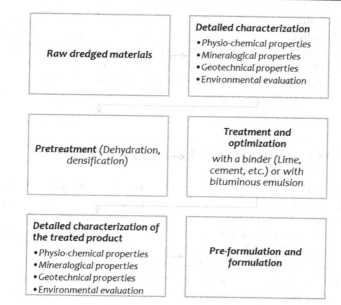

FIGURE 2.4   Approach for valorization of DMs in road construction.

2.2 USE OF SEDIMENTS FOR BRICKS AND CERAMIC PRODUCTION

Worldwide, reservoir sediments present environmental issues for all reservoirs. For this, mitigation methods like dredging and erosion have been proposed as general practices for environmental recovery. Several works have suggested different approaches for valorizing DMs in order to reduce pollution. DMs revealed different characteristics depending on their source. For example, sediments from seaports are constituted of sand as the main component, while silty particles present the majority in dam sediments and others [79]. The variation in the physio-chemical properties involved the need to add other wastes or raw materials such as slag and clays to manufacture construction materials such as bricks and ceramics [80, 81]. The PSD is the key parameter for ceramic and brick production [82]. Winkler (1954) [82] created a ternary diagram illustrating the distribution area of materials suitable for the manufacture of ceramic products and concluded that all DMs have a PSD suitable to produce full bricks without grinding [83]. DMs are mainly composed of quartz, illite, and geothite confirming the chemical content of silica (SiO_2), calcium, alumina, and iron oxide. Silica originates from quartz, feldspars, and muscovite. A high concentration of CaO indicates the presence of calcite ($CaCO_3$), common in marine environments, and presents the risk of efflorescence [84]. The existence of Al_2O_3 indicates the presence of clay phases [85]. In addition, Fe_2O_3 acts as fluxing mineral responsible for the color of the bricks after firing [83].

2.2.1 DMs from Fluvial and Lake Sources

Three river sediments were collected and mixed with fly ash (FA) in order to manufacture bricks and investigate their properties [86]. They showed a moderate plastic nature (> 18) and total fines of clays and silts ranged between 57% and 60%. The maximum strength of bricks revealed an unconfined compressive strength (UCS) of 27 N/mm², water absorption < 17% for 15 wt.% of FA, and firing at 1000 °C. In addition, polluted river DMs from France were stabilized through phosphatation (phosphate conversion coating) and used to produce bricks [87]. Treated DMs have been blended into bricks with 25%, 35%, and 45% by weight. It was proved that the plasticity decreased, and the water absorption increased by 10% by increasing the added sediments up to 45%. Furthermore, at this percentage, the compressive strength was about 26 MPa, which exceeded the values required by the standard (18–20 MPa).

The phosphatation process was also used to stabilize river DMs from Belgium [88]. This technology consists of adding phosphoric acid to the DMs containing calcite to form calcium phosphates and then proceeding with calcination to eliminate the OM. The used DMs are constituted by quartz, muscovite, anorthite, calcite, and hematite. It should be noted that quartz plays the role of sand, while the presence of hematite may enhance the flexural strength. The DMs in that study were incorporated to replace the sand by 15 wt.%. The manufactured bricks showed that adding 15% of treated DMs increased the compressive strength to 36 MPa and decreased the porosity to 15% in addition to the increase in firing shrinkage up to 40%. Furthermore, DMs from East Lake in China were mixed with different materials to produce ceramic facing bricks [89]. Several mixtures were manufactured using different amounts (by *wt.*) of some additives such as FA, feldspar, wollastonite, and coal gangue. The highest properties of 7.2% water absorption, porosity of ~16%, bulk density of 2.19 g/cm³, and bending strength of 46 MPa were obtained for 80% of East Lake sediment, with 10% of FA and 10% of wollastonite fired at 1100 °C. Savannah Harbor river DMs were studied as fully or 50% replacing natural clay for the manufacture of fired bricks using the extrusion process [90]. The bricks were fired at temperatures in a range of 900–1000 °C. The achieved compressive strength of 100% DMs bricks fired at 1000 °C was about 12 MPa, satisfying the standards for a low-weathering building brick. For the same firing temperature, using 50% of DMs increased the compressive strength to reach 29 MPa, thus in accordance with the ASTM requirements. Moreover, up to 50–70% of river sediments were blended with clays and then fired at 1100 °C to study their effect on the physio-mechanical properties of formulated bricks [91]. The findings revealed an increase of the bulk density from 1.21 to 1.45 g/cm^3. The high porosity led to the reduction of thermal conductivity by at least 40% for the fired bricks. Therefore, the maximum compressive strength of ~13 MPa was obtained for bricks with 50% of urban river DMs fired at 1050 °C. Three types of fired bricks were manufactured using lake sediments as the only raw material or mixed with cinder and sewage sludge [92]. It was proved that using 100% of lake sediments allows the optimal physio-mechanical properties. The compressive strength was ~33 MPa, with a bulk density of 1710 kg/m^3 and drying shrinkage of 6.5%.

River sediments were blended with cement, lime, FA, polymer, and jute fiber to create earth bricks with good mechanical properties [93]. The maximum compressive strength of 15 MPa was achieved for the brick's specimens, made of 65% of DMs, 25% of cement, 1.8% of polymer, and 1% of fibers. Additionally, DMs were employed for the production of permeable bricks [94]. The chemical analysis confirmed that DMs present a similar composition of natural clays used to produce bricks. In addition, leaching test (TCLP: Toxicity Characteristic Leaching Procedure) demonstrated that this material respects the standard requirements. The formulation of bricks consists of mixing the DMs with glass and ceramic waste at different proportions. The study results showed an important compressive strength of 159 MPa and porosity of 5.9% for bricks produced, using DMs to glass ratio of 1:3. River sediments and sewage sludge were used for the production of ceramics [95]. Contents of 15% sewage sludge and 5% of binder sintering at 1150 °C for 23 min were the optimum conditions to obtain the best properties. DMs were mixed with natural clays at 30%, 40%, and 50% [96]. Thermal conductivity of the bricks was in range of 0.45–0.50 W/mK for 30–50% of sediments. The reuse of DMs from Le Havre harbor (France) in fired bricks led to a bending strength of 16 MPa at firing temperature of 900 °C [83].

2.2.2 Marine Sediments

The study of DMs from Germany showed that their mineralogy, petrography, and chemistry make them suitable as raw material for brick manufacturing [97]. DMs were found to have a good capacity of solidification and stabilization of heavy metals during the production of bricks. In another study, dredging spoils, bottom ashes, sewage sludge, and steelworks slag were used as raw materials to produce ceramic tiles [98]. The obtained results revealed that a compressive strength of 79 MPa was reached for mixing spoils and sewage sludge at sintering temperature of 1150 °C. The density was ~2.94 g/cm³ and the water absorption was 1.1%. Polluted marine DMs were treated by the Novosol process, to a valorization in brick production [99]. Brick mixtures consisted of 60% of marine DMs to replace loam in natural clays and then were fired at 1100 °C. The mechanical behavior revealed a compressive strength of 38 MPa and water absorption coefficient of 9.3%. On the other hand, sintered ceramic was prepared using contaminated marine DMs to study the effect of the heating temperature on their properties [100]. At an optimum firing temperature of 1125 °C, the sintered ceramics revealed a bending strength of 22 MPa. The replacement of natural aggregates in concrete by sediments was between 12% and 15%, giving a compressive strength around 14 MPa [101]. Harbor DMs and slag waste were mixed and fired to produce non-hazardous construction materials [80]. The construction product type depends on firing temperature i.e., at a sintering temperature ≦ 1050 °C, the fired products are in a form of bricks, while at 1100 °C, they become lightweight aggregates. DMs are added with a substitution ratio of 19% for replacing quartz sand [102]. The study revealed that this substitution leads to a splitting tensile strength of 3.6 MPa and water absorption of 4.1%. Furthermore, the leaching test showed that the quantities of heavy metals were within acceptable limits. The characterization of fine-grained marine DMs showed a low plasticity silty clay [103]. Marine DMs were used to manufacture fired bricks prepared at a pressure of 20 MPa and fired at 1020 °C for 3.5 days. Physical and mechanical properties showed a compressive strength in the range between 40 and 46 MPa, and water absorption after boiling for 5 h of 12−14%, meeting the requirements.

The study of DMs' properties revealed their similarity to natural clays used in bricks [104]. The results showed a total shrinkage of 0.1%, an open porosity of 22%, bulk density of 2130 kg/m³, and vacuum water absorption of 10%. In addition, 50−60% of DMs from Brazilian seaports are mixed with demolition debris (20−35%) and lime wastes (15−30%) as compounds in composite materials [79]. The compressive strength reached 6.3 MPa at 3-d and 14.5 MPa at 90-d. The reported linear shrinkage was in range of 0.07−0.35%, while the water absorption was in range of 11.0−13.4%. Furthermore, contaminated DMs are employed to produce blocks by using binary cement and CO_2 curing to improve the compatibility of heavy metals and cement by the provision of sufficient magnesium hydrates [105]. One-day CO_2 curing transforms soluble magnesium hydrate to stable carbonates, densifying the microstructure and reducing the porosity from 18.5% to 16.9%. The addition of 5% of contaminated DMs could satisfy the strength requirements of the fill materials, showing a compressive strength of 2.7 MPa after 28-d [106]. Partition blocks made of 20% binders revealed a strength of sediment-based samples about 14 MPa at 7-d. Marine DMs are used as raw materials for the production of fired bricks and roofing tiles [107]. The products fired at 950 °C and 1050 °C have presented high water absorption of ~23%. While firing at 1100 °C, water absorption decreased to 7%, but these products tend to bend. The compressive and flexural strengths were 26 and 36 MPa when firing at 950 °C and 1050 °C, respectively. Bricks were manufactured with different amounts of sediments (up to 100 wt.%) and were fired at 850 °C and 950 °C [108]. The formulated bricks with 20% and 15% of DMs revealed an increase in strength of 33% and 21% for firing at 850 °C and 950 °C, respectively. However, the thermal conductivity of all bricks was low ranging from 0.24 to 0.43, and 0.21 to 0.46 W/mK for the bricks fired at 950 °C and 850 °C, respectively. Moreover, contaminated marine DMs and sewage sludge ash combined with cement/lime are used as construction materials [109]. The findings showed that the hardened samples based on 10% of lime and 20% of sewage sludge ash could reach a strength of 1.6 MPa after 28 days of curing.

2.2.3 Reservoir and Dam Sediments

Dam sediments and water treatment sludge were utilized as raw materials for fired-brick making [110]. The water absorption of dam sediments fired at 1050 °C was less than 15%, and bulk density and compressive strength were around 2.1 g/cm^3 and 600 kg/cm^2, respectively. Dam sediments fired at 1100 °C and above achieved a water absorption of ~1–3% and a compressive strength of above 400 kg/cm^2. Clays were added to reservoir sediments with a content up to 20% then sintered at temperatures of 1050 °C and 1100 °C to produce bricks [111]. The maximum density of 2.5 g/cm^3 was obtained for samples sintered at 1100 °C and 100% reservoir sediments. Dredging sediment soil was stabilized with cement for paving block, ceramic roof tile and concrete roof tile [112]. The amount of added cement was 5%, 10%, and 20% by weight as fine grain soil (silt-clay) that presented a plasticity index of 17%. The compressive strength increased from 0.35 to 14.5 kg/cm^2 for an amount of cement from 0 to 20%. In addition, reservoir sediment was employed to replace 100% of the soil content in conventional bricks [113]. While sodium bentonite, mica, FA, and bottom ash are used as pozzolan to substitute the cement in the sediment bricks, the optimum parameters were achieved for the proportion of 10% pozzolan (sodium bentonite, mica or FA). Nevertheless, 20% of pozzolans as cement replacement leads to a compressive strength that exceeds 13 MPa. Reservoir sediments are used as a replacement of the normal soils in the compressed bricks [114]. The water absorption increases, and the compressive strength decreased with increasing the content of added sediments. The optimum mix of 90%

sediments and 10% cement presented a compressive strength of 6.3 MPa and water absorption of 15%. In another study, the valorization of DMs in ceramic was investigated and resulted a flexural strength of 292 kgf/cm^2 with a density of 2.75 g/cm^3 for sintering temperature of 1140 °C [115].

2.2.4 Conclusions and Perspectives

The characterization of DMs is a key element in defining their suitability for fired bricks and ceramic production. Fired bricks have shown important potential for the use of DMs, presenting good properties. However, cementing options such as alkali-activation and geopolymerization, as compared to the firing method, present some advantages including a lower embodied energy method that minimizes carbon dioxide emission owing to the presence of clays that decompose at different temperatures. Nevertheless, methods to produce bricks are considerably dependent on raw materials including waste. As a result, this will positively affect economic and environmental aspects.

CREDIT AUTHORSHIP CONTRIBUTION STATEMENT

Safhi A.: Conceptualization, investigation, methodology, data curation, visualization, writing—original draft. **Mabroum S.:** Writing—original draft. **Bouchikhi A.:** Validation, writing—original draft. **Ez-Zaki H.:** Validation, writing—review and editing. **Benzaazoua M.:** Validation.

REFERENCES

1. C.L. Zhang, Q.S. Liu, J.B. Liu, Field plate load tests on dredged sediment dump pond with cement solidified crust above, *Adv. Mater. Res.* 255–260 (2011) 2751–2755. https://doi.org/10.4028/www.scientific.net/AMR.255-260.2751.
2. X. He, Y. Chen, Y. Wan, L. Liu, Q. Xue, Effect of curing stress on compression behavior of cement-treated dredged sediment, *Int. J. Geomech.* 20 (2020) 04020204. https://doi.org/10.1061/(ASCE)GM.1943-5622.0001857.
3. R. Zentar, V. Dubois, N.E. Abriak, Mechanical behaviour and environmental impacts of a test road built with marine dredged sediments, *Resour. Conserv. Recycl.* 52 (2008) 947–954. https://doi.org/10.1016/j.resconrec.2008.02.002.

4. V. Dubois, N.E. Abriak, R. Zentar, G. Ballivy, The use of marine sediments as a pavement base material, *Waste Manag.* 29 (2009) 774–782. https://doi.org/10.1016/j.wasman.2008.05.004.

5. R. Zentar, N. -E. Abriak, V. Dubois, M. Miraoui, Beneficial use of dredged sediments in public works, *Environ. Technol.* 30 (2009) 841–847. https://doi.org/10.1080/09593330902990139.

6. D.X. Wang, N.E. Abriak, R. Zentar, W. Xu, Solidification/stabilization of dredged marine sediments for road construction, *Environ. Technol.* 33 (2012) 95–101. https://doi.org/10.1080/09593330.2011.551840.

7. M.A. Bourabah, N. Abou-Bekr, S. Taibi, Geotechnical characterization of dredging sediments for valorization in road embankments: Case of the Cheurfas Dam (Algeria), in: *GeoFlorida 2010*, American Society of Civil Engineers, Orlando, FL, 2010: pp. 2212–2221. https://doi.org/10.1061/41095(365)224.

8. E. Silitonga, D. Levacher, S. Mezazigh, Utilization of fly ash for stabilization of marine dredged sediments, *Eur. J. Environ. Civ. Eng.* 14 (2010) 253–265. https://doi.org/10.1080/19648189.2010.9693216.

9. M. Miraoui, R. Zentar, N.-E. Abriak, Road material basis in dredged sediment and basic oxygen furnace steel slag, *Constr. Build. Mater.* 30 (2012) 309–319. https://doi.org/10.1016/j.conbuildmat.2011.11.032.

10. R. Zentar, D. Wang, N.E. Abriak, M. Benzerzour, W. Chen, Utilization of siliceous-aluminous fly ash and cement for solidification of marine sediments, *Constr. Build. Mater.* 35 (2012) 856–863. https://doi.org/10.1016/j.conbuildmat.2012.04.024.

11. D. Wang, N.E. Abriak, R. Zentar, Strength and deformation properties of Dunkirk marine sediments solidified with cement, lime and fly ash, *Eng. Geol.* 166 (2013) 90–99. https://doi.org/10.1016/j.enggeo.2013.09.007.

12. D. Wang, N.E. Abriak, R. Zentar, Co-valorisation of Dunkirk dredged sediments and siliceous–aluminous fly ash using lime, *Road Mater. Pavement Des.* 14 (2013) 415–431. https://doi.org/10.1080/14680629.2013.779309.

13. W. Maherzi, F.B. Abdelghani, Dredged Marine Sediments Geotechnical Characterisation for Their Reuse in Road Construction, *Eng. J.* 18 (2014) 27–37. https://doi.org/10.4186/ej.2014.18.4.27.

14. I.Y. Tang, D.Y.S. Yan, I.M.C. Lo, T. Liu, Pulverized fuel ash solidification/stabilization of waste: Comparison between beneficial reuse of contaminated marine mud and sediment, *J. Environ. Eng. Landsc. Manag.* 23 (2015) 202–210. https://doi.org/10.3846/16486897.2015.1021699.

15. D. Wang, N.E. Abriak, R. Zentar, Dredged marine sediments used as novel supply of filling materials for road construction, *Mar. Georesources Geotechnol.* 35 (2017) 472–480. https://doi.org/10.1080/1064119X.2016.1198945.

16. E. Silitonga, Experimental research of stabilization of polluted marine dredged sediments by using silica fume, *MATEC Web Conf.* 138 (2017) 01017. https://doi.org/10.1051/matecconf/201713801017.

17. M. Le Guern, T.A. Dang, M. Boutouil, Implementation and experimental monitoring of a subgrade road layer based on treated marine sediments, *J. Soils Sediments.* 17 (2017) 1815–1822. https://doi.org/10.1007/s11368-017-1652-1.

18. K. Kasmi, N.-E. Abriak, M. Benzerzour, H. Azrar, Environmental impact and mechanical behavior study of experimental road made with river sediments: Recycling of river sediments in road construction, *J. Mater. Cycles Waste Manag.* 19 (2017) 1405–1414. https://doi.org/10.1007/s10163-016-0529-5.

19. W. Maherzi, M. Benzerzour, Y. Mamindy-Pajany, E. van Veen, M. Boutouil, N.E. Abriak, Beneficial reuse of Brest-Harbor (France)-dredged sediment as alternative material in road building: Laboratory investigations, *Environ. Technol.* 39 (2018) 566–580. https://doi.org/10.1080/09593330.2017.1308440.

20. B. Serbah, N. Abou-Bekr, S. Bouchemella, J. Eid, S. Taibi, Dredged sediments valorisation in compressed earth blocks: Suction and water content effect on their mechanical properties, *Constr. Build. Mater.* 158 (2018) 503–515. https://doi.org/10.1016/j.conbuildmat.2017.10.043.

21. G. Fourvel, L. Vidal-Beaudet, A. Le Bocq, V. Brochier, F. Théry, D. Landry, T. Kumarasamy, P. Cannavo, Early structural stability of fine dam sediment in soil construction, *J. Soils Sediments.* 18 (2018) 2647–2663. https://doi.org/10.1007/s11368-018-1926-2.

22. F. Hamouche, R. Zentar, Effects of organic matter on mechanical properties of dredged sediments for beneficial use in road construction, *Environ. Technol.* 41 (2020) 296–308. https://doi.org/10.1080/09593330.2018.1497711.

23. T.-O. Ho, D.C.W. Tsang, W.-B. Chen, J.-H. Yin, Evaluating the environmental impact of contaminated sediment column stabilized by deep cement mixing, *Chemosphere.* 261 (2020) 127755. https://doi.org/10.1016/j.chemosphere.2020.127755.

24. A.P. Furlan, A. Razakamanantsoa, H. Ranaivomanana, O. Amiri, D. Levacher, D. Deneele, Effect of fly ash on microstructural and resistance characteristics of dredged sediment stabilized with lime and cement, *Constr. Build. Mater.* 272 (2021) 121637. https://doi.org/10.1016/j.conbuildmat.2020.121637.

25. R. Zentar, H. Wang, D. Wang, Comparative study of stabilization/solidification of dredged sediments with ordinary Portland cement and calcium sulfo-aluminate cement in the framework of valorization in road construction material, *Constr. Build. Mater.* 279 (2021) 122447. https://doi.org/10.1016/j.conbuildmat.2021.122447.

26. Z. Mkahal, Y. Mamindy-Pajany, W. Maherzii, N.-E. Abriak, Recycling of mineral solid wastes in backfill road materials: Technical and environmental investigations, *Waste Biomass Valorization.* (2021). https://doi.org/10.1007/s12649-021-01544-5.

27. A.B. Slama, N. Feki, D. Levacher, M. Zairi, Valorization of harbor dredged sediment activated with blast furnace slag in road layers, *Int. J. Sediment Res.* 36 (2021) 127–135. https://doi.org/10.1016/j.ijsrc.2020.08.001.

28. S. Bellara, M. Hidjeb, W. Maherzi, S. Mezazigh, A. Senouci, Optimization of an eco-friendly hydraulic road binders comprising clayey dam sediments and ground granulated blast-furnace slag, *Buildings.* 11 (2021) 443. https://doi.org/10.3390/buildings11100443.

29. O. Jan, B. Mir, Strength behaviour of cement stabilised dredged soil, *Int. J. Geosynth. Ground Eng.* 4 (2018). https://doi.org/10.1007/s40891-018-0133-y.

30. T. Tsuchida, A. Porbaha, N. Yamane, Development of a geomaterial from dredged bay mud, *J. Mater. Civ. Eng.* 13 (2001) 152–160. https://doi.org/10.1061/(ASCE)0899-1561(2001)13:2(152).

31. D.G. Wareham, J.R. Mackechnie, Solidification of New Zealand harbor sediments using cementitious materials, *J. Mater. Civ. Eng.* 18 (2006) 311–315. https://doi.org/10.1061/(ASCE)0899-1561(2006)18:2(311).

32. M. Salehi, N. Sivakugan, Effects of lime-clay modification on the consolidation behavior of the dredged mud, *J. Waterw. Port Coast. Ocean Eng.* 135 (2009) 251–258. https://doi.org/10.1061/(ASCE)WW.1943-5460.0000004.

33. L. Saussaye, E. van Veen, G. Rollinson, M. Boutouil, J. Andersen, J. Coggan, Geotechnical and mineralogical characterisations of marine-dredged sediments before and after stabilisation to optimise their use as a road material, *Environ. Technol.* 38 (2017) 3034–3046. https://doi.org/10.1080/09593330.2017,1287220.

34. D. Wang, R. Zentar, N.E. Abriak, Temperature-accelerated strength development in stabilized marine soils as road construction materials, *J. Mater. Civ. Eng.* 29 (2017) 04016281. https://doi.org/10.1061/(ASCE)MT.1943-5533.0001778.

35. M.O.A. Bazne, I.L. Howard, F. Vahedifard, Stabilized very high–moisture dredged soil: Relative behavior of portland-limestone cement and ordinary portland cement, *J. Mater. Civ. Eng.* 29 (2017) 04017110. https://doi.org/10.1061/(ASCE)MT.1943-5533.0001970.

36. M. Zelleg, I. Said, A. Missaoui, Z. Lafhaj, E. Hamdi, Dredged marine sediment as raw material in civil engineering applications, in: D.N. Singh, A. Galaa (Eds.), *Contemp. Issues Geoenvironmental Eng.*, Springer International Publishing, Cham, 2018: pp. 407–418. https://doi.org/10.1007/978-3-319-61612-4_33.

37. M. Zelleg, I. Said, E. Hamdi, Z. Lafhaj, Experimental testing for Zarzis port sediments (Tunisia) in road materials, *Geotech. Res.* 5 (2018) 13–21. https://doi.org/10.1680/jgere.17.00013.

38. T.T.M. Nguyen, S. Rabbanifar, N.A. Brake, Q. Qian, K. Kibodeaux, H.E. Crochet, S. Oruji, R. Whitt, J. Farrow, B. Belaire, P. Bernazzani, M. Jao, Stabilization of silty clayey dredged material, *J. Mater. Civ. Eng.* 30 (2018) 04018199. https://doi.org/10.1061/(ASCE)MT.1943-5533.0002391.

39. D. Wang, R. Zentar, N.E. Abriak, Durability and swelling of solidified/stabilized dredged marine soils with class-F fly ash, cement, and lime, *J. Mater. Civ. Eng.* 30 (2018) 04018013. https://doi.org/10.1061/(ASCE)MT.1943-5533.0002187.

40. H. Lei, Y. Xu, X. Li, G. Zheng, G. Liu, Effects of polyacrylamide on the consolidation behavior of dredged clay, *J. Mater. Civ. Eng.* 30 (2018) 04018022. https://doi.org/10.1061/(ASCE)MT.1943-5533.0002201.

41. K.H. Khayat, J. Assaad, J. Daczko, Comparison of field-oriented test methods to assess dynamic stability of self-consolidating concrete, *ACI Mater. J.* 101 (2004). https://doi.org/10.14359/13066.

42. D. Wang, J. Xiao, X. Gao, Strength gain and microstructure of carbonated reactive MgO-fly ash solidified sludge from East Lake, China, *Eng. Geol.* 251 (2019) 37–47. https://doi.org/10.1016/j.enggeo.2019.02.012.

43. J. Li, Y. Zhou, Q. Wang, Q. Xue, C.S. Poon, Development of a novel binder using lime and incinerated sewage sludge ash to stabilize and solidify contaminated marine sediments with high water content as a fill material, *J. Mater. Civ. Eng.* 31 (2019) 04019245. https://doi.org/10.1061/(ASCE)MT.1943-5533.0002913.

44. Y. Liu, D.E.L. Ong, E. Oh, Z. Liu, R. Hughes, Sustainable cementitious blends for strength enhancement of dredged mud in Queensland, Australia, *Geotech. Res.* (2022) 1–18. https://doi.org/10.1680/jgere.21.00046.

45. T. Chompoorat, T. Thepumong, S. Taesinlapachai, S. Likitlersuang, Repurposing of stabilised dredged lakebed sediment in road base construction, *J. Soils Sediments.* 21 (2021) 2719–2730. https://doi.org/10.1007/s11368-021-02974-3.

46. T. Chompoorat, S. Likitlersuang, T. Thepumong, W. Tanapalungkorn, P. Jamsawang, P. Jongpradist, Solidification of sediments deposited in reservoirs with cement and fly ash for road construction, *Int. J. Geosynth. Ground Eng.* 7 (2021) 85. https://doi.org/10.1007/s40891-021-00328-0.

47. T. Chompoorat, K. Thanawong, S. Likitlersuang, Swell-shrink behaviour of cement with fly ash-stabilised lakebed sediment, *Bull. Eng. Geol. Environ.* 80 (2021) 2617–2628. https://doi.org/10.1007/s10064-020-02069-2.

48. Maher, T. Bennert, F. Jafari, W.S. Douglas, N. Gucunski, Geotechnical properties of stabilized dredged material from New York-New Jersey harbor, *Transp. Res. Rec.* 1874 (2004) 86–96. https://doi.org/10.3141/1874-10.

49. D.X. Wang, H.X. Wang, W.Z. Chen, Reinforcement mechanism of cement/lime-fly ash treated sediments as road construction materials, *Appl. Mech. Mater.* 99–100 (2011) 924–927. https://doi.org/10.4028/www.scientific.net/AMM.99-100.924.

50. W. Zoubir, K. Harichane, M. Ghrici, Effect of lime and natural pozzolana on dredged sludge engineering properties, *Electron. J. Geotech. Eng.* 18 C (2013) 589–600.

51. Azhar, C.M. Chan, strength development in cement admixed fine-grained dredged marine soils, *Appl. Mech. Mater.* 802 (2015) 272–276. https://doi.org/10.4028/www.scientific.net/AMM.802.272.

52. L. Larouci, S. Yassine, L. Laïd, B. Amar, Improvement of the mechanical performance of Fergoug dam sediments treated for reuse in road engineering, *MATEC Web Conf.* 149 (2018) 01031. https://doi.org/10.1051/matecconf/201814901031.

53. N. Yoobanpot, P. Jamsawang, H. Poorahong, P. Jongpradist, S. Likitlersuang, Multiscale laboratory investigation of the mechanical and microstructural properties of dredged sediments stabilized with cement and fly ash, *Eng. Geol.* 267 (2020) 105491. https://doi.org/10.1016/j.enggeo.2020.105491.

54. P. Jamsawang, S. Charoensil, T. Namjan, P. Jongpradist, S. Likitlersuang, Mechanical and microstructural properties of dredged sediments treated with cement and fly ash for use as road materials, *Road Mater. Pavement Des.* 0 (2020) 1–25. https://doi.org/10.1080/14680629.2020.1772349.

55. H. Poorahong, N. Wongvatana, P. Jamsawang, K. Lueprasert, K. Tantayopin, X.B. Chen, Unconfined compressive and splitting tensile strength of dredged sediments stabilized with cement and fly ash, *Key Eng. Mater.* 856 (2020) 367–375. https://doi.org/10.4028/www.scientific.net/KEM.856.367.

56. A. Larouci, Y. Senhadji, L. Laoufi, A. Benazzouk, Valorisation of natural waste: Dam sludge for road construction, *Nat. Environ. Pollut. Technol.* 19 (2020) 1075–1083. https://doi.org/10.46488/NEPT.2020.v19i03.018.

57. N. Gueffaf, B. Rabehi, K. Boumchedda, Recycling dam sediments for the elaboration of stabilized blocks, *Int. J. Eng. Res. Afr.* 50 (2020) 131–144. https://doi.org/10.4028/www.scientific.net/JERA.50.131.

58. X. Cheng, Y. Chen, G. Chen, B. Li, Characterization and prediction for the strength development of cement stabilized dredged sediment, *Mar. Georesources Geotechnol.* 39 (2021) 1015–1024. https://doi.org/10.1080/1064119X.2020.1795014.

59. L. Lang, B. Chen, D. Li, Effect of nano-modification and fiber-reinforcement on mechanical behavior of cement-stabilized dredged sediment, *Mar. Georesources Geotechnol.* 0 (2021) 1–17. https://doi.org/10.1080/1064119X.2021.1954112.

60. Y. Zhou, G. Cai, C. Cheeseman, J. Li, C.S. Poon, Sewage sludge ash-incorporated stabilisation/solidification for recycling and remediation of marine sediments, *J. Environ. Manage.* 301 (2022) 113877. https://doi.org/10.1016/j.jenvman.2021.113877.

61. D.G. Grubb, N.E. Malasavage, C.J. Smith, M. Chrysochoou, Stabilized dredged material. II: Geomechanical behavior, *J. Geotech. Geoenvironmental Eng.* 136 (2010) 1025–1036. https://doi.org/10.1061/(ASCE)GT.1943-5606.0000290.

62. Y.T. Kim, J. Ahn, W.J. Han, M.A. Gabr, Experimental evaluation of strength characteristics of stabilized dredged soil, *J. Mater. Civ. Eng.* 22 (2010) 539–544. https://doi.org/10.1061/(ASCE)MT.1943-5533.0000052.

63. N. Kasmi, N.-E. Abriak, M. Benzerzour, H. Azrar, Effect of dewatering by the addition of flocculation aid on treated river sediments for valorization in road construction, *Waste Biomass Valorization.* 8 (2017) 585–597. https://doi.org/10.1007/s12649-016-9587-0.

64. H. Yu, J. Yin, A. Soleimanbeigi, W.J. Likos, Effects of curing time and fly ash content on properties of stabilized dredged material, *J. Mater. Civ. Eng.* 29 (2017) 04017199. https://doi.org/10.1061/(ASCE)MT.1943-5533.0002032.

65. S. Mohammad, W. Akram, S.A. Mirza, Geotechnical characterization of dredged material and effect of lime stabilisation on its strength characteristics, *Appl. Mech. Mater.* 877 (2018) 289–293. https://doi.org/10.4028/www.scientific.net/AMM.877.289.

66. B.T. Smith, I.L. Howard, F. Vahedifard, Lightly cemented dredged sediments for sustainable reuse, *Environ. Geotech.* 5 (2018) 324–335. https://doi.org/10.1680/jenge.16.00019.

67. H. Huang, S.E. Burns, K.E. Kurtis, Beneficial use of Savannah River dredged material in large-scale geotechnical applications, *Jpn. Geotech. Soc. Spec. Publ.* 9 (2021) 245–248. https://doi.org/10.3208/jgsssp.v09.cpeg064.

68. M. Houlihan, G. Bilgen, A.Y. Dayioglu, A.H. Aydilek, Geoenvironmental evaluation of RCA-stabilized dredged marine sediments as embankment material, *J. Mater. Civ. Eng.* 33 (2021) 04020435. https://doi.org/10.1061/(ASCE)MT.1943-5533.0003547.

69. L. Lang, B. Chen, Stabilization of dredged sediment using activated binary cement incorporating nanoparticles, *J. Mater. Civ. Eng.* 34 (2022) 04021381. https://doi.org/10.1061/(ASCE)MT.1943-5533.0004017.

70. A. El-Shinawi, V. Kramarenko, Assessment of stabilized dredged sediments using portland cement for geotechnical engineering applications along Hurghada Coast, Red Sea, Egypt, *Asian J. Appl. Sci.* 3 (2015). https://192.99.73.24/index.php/AJAS/article/view/3166 (accessed March 29, 2022).

71. G. Kang, T. Tsuchida, T.X. Tang, T.P. Kalim, Consistency measurement of cement-treated marine clay using fall cone test and Casagrande liquid limit test, *Soils Found.* 57 (2017) 802–814. https://doi.org/10.1016/j.sandf.2017.08.010.

72. M.Z. Rosman, C.-M. Chan, N.M. Anuar, Compressibility and permeability of solidified dredged marine soils (DMS) with the addition of cement and/or waste granular materials (WGM), *J. Sci. Technol.* 10 (2018). https://publisher.uthm.edu.my/ojs/index.php/JST/article/view/3648 (accessed April 11, 2022).

73. M. Mustin, *Le compost: gestion de la matière organique*, (1987).

74. L.A. Paassen, L.F. Gareau, Effect of pore fluid salinity on compressibility and shear strength development of clayey soils, (2004) 327–340. https://doi.org/10.1007/978-3-540-39918-6_39.

75. E. Rozière, M. Samara, A. Loukili, D. Damidot, Valorisation of sediments in self-consolidating concrete: Mix-design and microstructure, *Constr. Build. Mater.* 81 (2015) 1–10. https://doi.org/10.1016/j.conbuildmat.2015.01.080.

76. A. Bouchikhi, *Optimisation de la valorisation des déchets de verre et de sédiments dans des liants recomposés: Activation - Formulation de mortiers - Stabilisation physico-chimique*. Génie civil. Ecole nationale supérieure Mines-Télécom Lille Douai, Français. (2020). https://www.theses.fr/2020MTLD0018.

77. V. Dubois, R. Zentar, N.-E. Abriak, P. Grégoire, Fine sediments as a granular source for civil engineering, *Eur. J. Environ. Civ. Eng.* 15 (2011) 137–166. https://doi.org/10.1080/19648189.2011.9693315.

78. W. Maherzi, *Sustainable environmental treatment and reuse of dredged marine sediments in road construction*, World Congress of International Solid Waste Association (ISWA), Daegu, Korea, (2010).

79. V. Mymrin, J.C. Stella, C.B. Scremim, R.C.Y. Pan, F.G. Sanches, K. Alekseev, D.E. Pedroso, A. Molinetti, O.M. Fortini, Utilization of sediments dredged from marine ports as a principal component of composite material, *J. Clean. Prod.* 142 (2017) 4041–4049. https://doi.org/10.1016/J.JCLEPRO.2016.10.035.

80. Y.L. Wei, C.Y. Lin, S.H. Cheng, H.P. Wang, Recycling steel-manufacturing slag and harbor sediment into construction materials, *J. Hazard. Mater.* 265 (2014) 253–260. https://doi.org/10.1016/J.JHAZMAT.2013.11.049.

81. F. Haurine, I. Cojan, M.A. Bruneaux, Development of an industrial mineralogical framework to evaluate mixtures from reservoir sediments for recovery by the heavy clay industry: Application of the Durance system (France), *Appl. Clay Sci.* 132–133 (2016) 508–517. https://doi.org/10.1016/J.CLAY.2016.07.022.

82. M. Dondi, B. Fabbri, G. Guarini, Grain-size distribution of Italian raw materials for building clay products: A reappraisal of the Winkler diagram, *Clay Miner.* 33 (1998) 435–442. https://doi.org/10.1180/000985598545732.

83. L. Mesrar, A. Benamar, B. Duchemin, S. Brasselet, F. Bourdin, R. Jabrane, Engineering properties of dredged sediments as a raw resource for fired bricks, *Bull. Eng. Geol. Environ.* 80 (2021) 2643–2658. https://doi.org/10.1007/S10064-020-02068-3/TABLES/6.

84. M. Dondi, B. Fabbri, G. Guarini, M. Marsigli, C. Mingazzini, Soluble salts and efflorescence in structural clay products: A scheme to predict the risk of efflorescence, *Bol Soc Esp Ceram Vidr.* 36 (1997) 619–629.

85. B.S. Nabawy, N.T.H. Elgendy, M.T. Gazia, Mineralogic and diagenetic controls on reservoir quality of paleozoic sandstones, Gebel El-Zeit, North Eastern Desert, Egypt, *Nat. Resour. Res.* 29 (2020) 1215–1238. https://doi.org/10.1007/S11053-019-09487-4/FIGURES/14.

86. J.M. Bhatnagar, R.K. Goel, R.G. Gupta, Brick-making characteristics of river sediments of the South West Bengal region of India, *Constr. Build. Mater.* 8 (1994) 177–183. https://doi.org/10.1016/S0950-0618(09)90032-0.

87. Z. Lafhaj, M. Samara, F. Agostini, L. Boucard, F. Skoczylas, G. Depelsenaire, Polluted river sediments from the North region of France: Treatment with Novosol® process and valorization in clay bricks, *Constr. Build. Mater.* 22 (2008) 755–762. https://doi.org/10.1016/J.CONBUILDMAT.2007.01.023.

88. M. Samara, Z. Lafhaj, C. Chapiseau, Valorization of stabilized river sediments in fired clay bricks: Factory scale experiment, *J. Hazard. Mater.* 163 (2009) 701–710. https://doi.org/10.1016/J.JHAZMAT.2008.07.153.

89. J. Wu, G. Leng, X. Xu, Y. Zhang, X. Lao, K. Li, Preparation and properties of ceramic facing brick from East-lake sediment, *J. Wuhan Univ. Technol.-Mater Sci Ed.* 27 (2012) 154–159.

90. A. Mezencevova, N.N. Yeboah, S.E. Burns, L.F. Kahn, K.E. Kurtis, Utilization of Savannah Harbor river sediment as the primary raw material in production of fired brick, *J. Environ. Manage.* 113 (2012) 128–136. https://doi.org/10.1016/J .JENVMAN.2012.08.030.

91. Y. Xu, C. Yan, B. Xu, X. Ruan, Z. Wei, The use of urban river sediments as a primary raw material in the production of highly insulating brick, *Ceram. Int.* 40 (2014) 8833–8840. https://doi.org/10.1016/J.CERAMINT.2014.01.105.

92. Y.M. Zhang, L.T. Jia, H. Mei, Q. Cui, P.G. Zhang, Z.M. Sun, Fabrication, micro-structure and properties of bricks fired from lake sediment, cinder and sew-age sludge, *Constr. Build. Mater.* 121 (2016) 154–160. https://doi.org/10.1016/J .CONBUILDMAT.2016.05.155.

93. J.X. Liu, R. Hai, L. Zhang, Modification of Yellow River sediment based sta-bilized earth bricks, *Kem. U Ind. Časopis Kemičara Kem. Inženjera Hrvat.* 65 (2016) 613–618. https://doi.org/10.15255/KUI.2016.023.

94. C. Zhou, X. Cheng, L. Zeng, H. Wang, J. Chen, Feasibility of using dredged mud for prepared the permeable brick, *IOP Conf. Ser. Mater. Sci. Eng.* 250 (2017) 012030. https://doi.org/10.1088/1757-899X/250/1/012030.

95. Y. Jin, S. Huang, Q. Wang, M. Gao, H. Ma, Ceramsite production from sediment in Beian River: Characterization and parameter optimization, *R. Soc. Open Sci.* 6 (2019). https://doi.org/10.1098/RSOS.190197.

96. M. Manni, F.F. de Albuquerque Landi, T. Giannoni, A. Petrozzi, A. Nicolini, F. Cotana, A comparative study on opto-thermal properties of natural clay bricks incorporating dredged sediments, *Energies* 14 (2021) 4575. https://doi.org/10 .3390/EN14154575.

97. K. Hamer, V. Karius, Brick production with dredged harbour sediments. An industrial-scale experiment, *Waste Manag.* 22 (2002) 521–530. https://doi.org/10 .1016/S0956-053X(01)00048-4.

98. D. Baruzzo, D. Minichelli, S. Bruckner, L. Fedrizzi, A. Bachiorrini, S. Maschio, Possible production of ceramic tiles from marine dredging spoils alone and mixed with other waste materials, *J. Hazard. Mater.* 134 (2006) 202–210. https:// doi.org/10.1016/J.JHAZMAT.2005.10.053.

99. L. Zoubeir, S. Adeline, C.S. Laurent, C. Yoann, H.T. Truc, L.G. Benoît, A. Federico, The use of the Novosol process for the treatment of polluted marine sediment, *J. Hazard. Mater.* 148 (2007) 606–612. https://doi.org/10.1016/J .JHAZMAT.2007.03.029.

100. M. Romero, A. Andrés, R. Alonso, J. Viguri, J.M. Rincón, Sintering behaviour of ceramic bodies from contaminated marine sediments, *Ceram. Int.* 34 (2008) 1917–1924. https://doi.org/10.1016/J.CERAMINT.2007.07.002.

101. M. Zdiri, N. Abriak, M. Ben Ouezdou, J. Neji, The use of fluvial and marine sediments in the formulation of Roller Compacted Concrete for use in pave-ments, *Environ. Technol.* 30 (2009) 809–815. https://doi.org/10.1080/09593330 902990097.

102. A. Said, A. Missaoui, Z. Lafhaj, Reuse of Tunisian marine sediments in paving blocks: Factory scale experiment, *J. Clean. Prod.* 102 (2015) 66–77. https://doi. org/10.1016/J.JCLEPRO.2015.04.138.

103. I.M.G. Bertelsen, L.J. Belmonte, W. Chen, L.M. Ottosen, Properties of bricks produced from Greenlandic marine sediments, in: Proceedings 23rd International Conference on Port and Ocean Engineering under Arctic Conditions (POAC '15), Trondheim, Norway. (2015).

104. L.J. Belmonte, I.M.G. Bertelsen, Evaluation of the potential for using Greenlandic marine sediments for brick production. In International Conference on Materials, Systems and Structures in Civil Engineering: Conference workshop on Cold Region Engineering Technical University of Denmark, Kongens Lyngby, Danemark, (2016) 1–6.

105. L. Wang, T.L.K. Yeung, A.Y.T. Lau, D.C.W. Tsang, C.S. Poon, Recycling contaminated sediment into eco-friendly paving blocks by a combination of binary cement and carbon dioxide curing, *J. Clean. Prod.* 164 (2017) 1279–1288. https://doi.org/10.1016/J.JCLEPRO.2017.07.070.

106. L. Wang, L. Chen, D.C.W. Tsang, J.S. Li, K. Baek, D. Hou, S. Ding, C.S. Poon, Recycling dredged sediment into fill materials, partition blocks, and paving blocks: Technical and economic assessment, *J. Clean. Prod.* 199 (2018) 69–76. https://doi.org/10.1016/J.JCLEPRO.2018.07.165.

107. P. Baksa, F. Cepak, R. Kovačič Lukman, V. Ducman, An evaluation of marine sediments in terms of their usability in the brick industry: Case study port of koper, *J. Sustain. Dev. Energy Water Environ. Syst.* 6 (2018) 78–88. https://doi .org/10.13044/J.SDEWES.D5.0183.

108. H. Slimanou, D. Eliche-Quesada, S. Kherbache, N. Bouzidi, A. /K Tahakourt, Harbor Dredged Sediment as raw material in fired clay brick production: Characterization and properties, *J. Build. Eng.* 28 (2020) 101085. https://doi.org/ 10.1016/J.JOBE.2019.101085.

109. J.S. Li, Y. Zhou, X. Chen, Q. Wang, Q. Xue, D.C.W. Tsang, C.S. Poon, Engineering and microstructure properties of contaminated marine sediments solidified by high content of incinerated sewage sludge ash, *J. Rock Mech. Geotech. Eng.* 13 (2021) 643–652. https://doi.org/10.1016/J.JRMGE.2020.10.002.

110. C. Huang, J.R. Pan, K.D. Sun, C.T. Liaw, Reuse of water treatment plant sludge and dam sediment in brick-making, *Water Sci. Technol.* 44 (2001) 273–277. https://doi.org/10.2166/WST.2001.0639.

111. K.Y. Chiang, K.L. Chien, S.J. Hwang, Study on the characteristics of building bricks produced from reservoir sediment, *J. Hazard. Mater.* 159 (2008) 499–504. https://doi.org/10.1016/J.JHAZMAT.2008.02.046.

112. H. Yusuf, M.S. Pallu, L. Samang, M.W. Tjaronge, The utilization of Bili-bili Dam's dredging sediment stabilized with cement for construction material, *Int. J. Civ. Environ. Eng.* 12(4) (2012), 25–31.

113. H.Q. Chua, L.W. Ean, B.S. Mohammed, M.A. Malek, L.S. Wong, C.W. Tang, Z.K. Lee, A.F. Lim, Y.Y. See, J.L. Ooi, *Potential Use of Cameron Highlands Reservoir Sediment in Compressed Bricks*, Key Engineering Materials, 594–595 2013, 487–491.

114. J.L. Ooi, L.W. Ean, B.S. Mohammed, M.A. Malek, L.S. Wong, C.W. Tang, H.Q. Chua, Study on the properties of compressed bricks using Cameron highlands reservoir sediment as primary material, *Appl. Mech. Mater.* 710 (2015) 25–29. https://doi.org/10.4028/WWW.SCIENTIFIC.NET/AMM.710.25.

115. L.K. Kreirzti, L. Benamara, N.E. Boudjenane, Valorization of dredging sediments of dam BOUHNIFIA in ceramic, *J. Aust. Ceram. Soc.* 55 (2019) 1081–1089. https://doi.org/10.1007/S41779-019-00321-X/FIGURES/9.

Reuse of Dredged Materials as Filler, Fine, and Lightweight Aggregates in Cementitious Matrix

3

Amine El Mahdi Safhi, Hassan Ez-Zaki,
Patrice Rivard, Mostafa Benzaazoua

Contents

DOI: 10.1201/9781003315551-3

43

ABSTRACT

Several studies have been conducted on recycling dredged materials (DMs), clayey and sandy, as filler, fine aggregate, and as lightweight aggregates (LWA), incorporated in concrete. Recent research confirms the potential beneficial use of DMs as filler; the results suggest a substitution rate of the fine aggregates by DMs up to 20–50% without altering the mechanical performance. The main concern with LWA produced from DMs is their higher water absorption capacity as well as their organic matter content as blowing agent, which decreases the density and mechanical strength, and increases the porosity and the penetrability of chlorides. The existing studies reported a wide range in the specific gravity of 0.88–2.51, a bulk density of 544–1500 kg/m^3, and a water absorption of 0.6–51%. Overall, in-depth research is required to confirm such valorization pathways despite the variability of DMs based on their provenance, physiochemical properties, and their contaminating potential.

3.1 VALORIZATION OF DREDGED MATERIALS AS A FILLER

Only six publications were found on the topic of reusing sediments as filler. The main purpose of utilizing DMs as filler is to minimize the consumption of natural raw materials and the environmental risks. Several studies have proved the possibility of reusing these materials as filler completely or partly in concretes [1–6]. In Singapore, using DMs as filler in ultra-high-performance concrete (UHPC) proved to be an efficient means to mitigate the overall carbon footprint of the concrete and make it sustainable and eco-friendly [3]. Located DMs are considered as having low-grade kaolinite clay with about 30% kaolinite content. The DMs were treated by drying, grinding, and calcination for one hour at 700 °C. The partial replacement (up to 30 wt.%) of the costly quartz powder by calcined sediments in UHPC has shown a comparable strength with the control concrete (> 120 MPa). Moreover, the examination of the hydration kinetics and degree of hydration showed a considerable acceleration when the level of replacement was increased. Rozas *et al.* 2015 [1] have published guidelines on the assessment of marine DMs' valorization in self-consolidating concrete (SCC) and proved the feasibility of the usage of DMs for the production of structural cementitious materials. Additionally, the properties of concrete fresh state based on DMs as fillers showed a cohesive and self-leveling property similar to SCC [1]. Other studies [5], however, have mentioned that untreated marine DMs have increased the yield stress of SCC due to the fineness of the particles and the presence of organic matter. On the other hand, dried raw DMs were not recommended for the SCC formulation because the agglomeration can reduce the workability and the mechanical properties after mixing [7]. Rozière *et al.* 2015 [2] have highlighted the beneficial utilization of treated DMs as substitute of limestone filler and aggregates in SCC. Treated sediment has significantly improved the compressive strength of mortars. Moreover, DM-based concrete showed comparable strength to that of the control. Nevertheless, treated sediment has increased the yield stress and plastic viscosity and thus decreased the slump flow and workability of concrete. Therefore, treated sediment can negatively influence the fresh state properties of SCC due to its high internal porosity which tends to adsorb a portion of the mixing water resulting in the flow resistance [2]. Xu *et al.* 2020 [8] blended a series of ordinary Portland cement (OPC) with DMs (as fillers) and mixed with sodium chloride solutions (different concentrations) for evaluation of the effect of salinity on the rheological properties of fresh pastes. The slump and rheology tests exhibited an increase in the slump flow and a decrease of the rheological properties i.e., yield stress and viscosity, with increasing salinity. Moreover, increasing salinity of the solution can limit the mixing water in the binders causing lower viscosity for the same water-to-binder (*w/b*) ratio. Therefore, it was suggested that the use of marine DMs can increase the workability and reduce the liquid limit of the binders when used untreated, taking into consideration other impacts of the salinity on the physio-chemical and mechanical properties, and durability.

A recent study by Beddaa *et al.* 2022 [6] revealed the beneficial use of fine DMs as fillers compared to their use as fine aggregates in concrete. It is shown that the replacement of sand with 30% of fine DMs can delay the hydration kinetics, reduce compressive strength, and increase shrinkage of the concrete. Regarding OPC replacement with only 10% of fine DMs, the findings exhibited acceptable technical, economic, and environmental benefits, as illustrated in Figure 3.1.

Up to 10% substitution rate of fine DMs as cement replacement did not significantly affect the concrete properties such as workability, hydration kinetics (~3h of delay) and compressive strength (~8% reduction). On the other

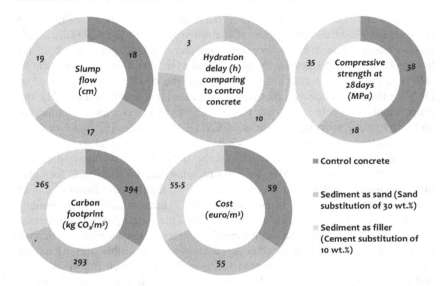

FIGURE 3.1 Concrete properties, carbon footprint, and cost of different concretes based on fine DMs as filler or as sand. Constructed from Buildings, 12(2), Beddaa H. *et al.* "Reuse of untreated fine sediments as filler: is it more beneficial than incorporating them as sand?", 211. Copyright 2022, Open Access.

hand, a reduction in the carbon footprint and cost of concrete with sediment as filler was acquired. Approximately 10% of CO_2 was reduced from concrete with sediment as filler, if only 10% of OPC was replaced by fine DMs compared to concrete with 30% of sediment as sand (~0.3% of CO_2 reduction). The cost of using 10% of fine DMs as filler can save about 6% of the ordinary concrete cost due to the replacement of cement, which is considered the most expensive ingredient.

3.2 VALORIZATION OF DREDGED MATERIALS AS FINE AGGREGATES

Thirty-one papers (11% of the database) have reported studies carried out on recycling DMs as fine aggregates in concrete. This is a critical upgrade considering the important impact of fine aggregates on several properties of concrete including workability, packing density, mechanical strength, etc. The first paper was published by Agostini *et al.* in 2007 [9]. The authors investigated the feasibility of polluted treated DMs from France. The mineralogical composition of raw DMs endorses their use as fine aggregate: 5% of clay (< 2 μm), 59% of silt (2–63 μm), and 36% of sand (> 63 μm) [9]. Some studies reported only a detailed characterization for such use [10], while others reported the effect of DMs on different properties of the matrix.

3.2.1 Effects of DMs on Mechanical Strength

The most significant parameter to determine the direct effect of the addition in the cementitious matrix is to evaluate the compressive strength of the mixture. Among other parameters are the Flexural (bending) strength, associated with the modulus of rupture, generally derived by applying a load on three or four points, and the tensile splitting strength, a property that shapes the pattern of the matrix exposed to torsion and transverse shear. Table 3.1 summarizes the effects of DMs on compressive strength reported in 22 studies. From the reported effects, the replacement level of 20–50% was found to have a comparable compressive strength. Even for the total replacement, some studies reported a slight decrease in strength up to 20% compared to the control. Table 3.2 describes the effect of DMs on the flexural and splitting tensile strengths. Also, a range of 20–50% incorporation rate was found to have strengths comparable to the control. However, two studies have reported a decrease in the mechanical performance for mortars based on a DM quantity of lower than 20%. Several parameters linked to the fine aggregates could affect the compressive strength including the additional water linked to the high capillarity nature of those materials, the grading, the morphology, and most of all their chemical composition and solubility, e.g., content of chloride and sulfate. For an adapted chemical stability, some studies suggested a pretreatment to minimize those aggressive contents (washing DMs by applying a treatment process e.g., Novosol®).

3.2.2 Effects of DMs on Young's Modulus

Young's modulus (E) assesses the linear stress-strain relation when subjected to vertical load in the elastic spectrum i.e., slope ratio of the linear section relationship [32]. According to Hooke's law, the elastic limit is a prominent material stress capable of maintaining, without deviating from the proportionality, stress to strain. Only six studies have reported the effect of DMs on (E). It was reported that (E) remains stable up to 33% substitution, but above this rate it regularly decreases with increasing DMs amount [9]. Another study has reported that (E) of concretes made with DMs (up to 50%) was similar to those

TABLE 3.1 Effect of DMs on Compressive Strength

REF.	REPLACEMENT RATE	MIX	EFFECT ON COMPRESSIVE STRENGTH
[9, 11]	0, 33, 66, and 100%	Mortar	33% was the optimal substitution ratio; above this, the strength decreased but was still comparable with the control.
[12]	0, 5, 10, 30, and 70%	Mortar	The strength increased when the substitution rate of DMs was used up to 39%.
[13–15]	0, 15, 25, 35, and 50%	Mortar and concrete	Similar compressive strengths were obtained.
[16]	0, 25, 50, 75, and 100%	Mortar	The formulations based on 25–50% of DMs had a comparable strength.
[17]	0, 10, 15, and 20%	Mortar	At 90 days, all mortars made with DMs were close to those of the control.
[18]	0 and 20%	Mortar	Strength of concrete mix prepared with 20 wt.% was comparable to that of the control mixture.
[19]	0 and 100%	Mortar	Final strength formulated with total DMs hardly reached 20% of the final strength of control.
[20]	0, 10, 20, 30, 40, and 50%	Mortar	The strength of concrete made with DMs showed a decreasing strength as the DMs content increases.
[21]	0, 25, 50, 75, and 100%	Concrete	Strength decreased with an increase in DM content. A pre-treatment was required to reduce Cl^- and SO_4^{2-} contents.
[22]	20%	Mortar	Strength decreased with increase in DM content.
[23]	12.5 and 20%	Concrete	Strength decreased with increase in the DM content.
[24]	0 and 100%	Concrete	A slight decrease of compressive strength was noticed.
[25, 26]	50, 70, and 100%	Concrete	Corrected granulometry by adding 2.5/10 mm grain gave much better compressive strength.

(Continued)

TABLE 3.1 (CONTINUED) Effect of DMs on Compressive Strength

REF.	REPLACEMENT RATE	MIX	EFFECT ON COMPRESSIVE STRENGTH
[27]	0 and 30%	Concrete	Slight decrease when substituting the coarser sand, and up to 50% decrease when replacing the fine fraction of the sand.
[28]	0 to 100%	Concrete	25% DM-based concrete had a comparable strength at 60 and 90-d. The concretes were cured at 150, 350, and 600 °C.
[7]	30%	Mortar	The mortars reached around 38 MPa at 90-d, no control used.
[29]	100%	Concrete	The strengths of concrete-based sediment were similar to that of the control concrete or slightly lower.
[30]	100%	Concrete	Concretes exhibit lower early-age and long-term compressive strengths.
[31]	0 to 100%	Mortar	Increasing the incorporation of the dredged sand decreased the compressive strength.

of the control concrete [13–15]. Ozer-Erdogan *et al.* (2016) [21] founded a comparable modulus with a substitution rate up to 100%. However, in another study with a low substitution rate up to 20%, Achour *et al.* (2019) [23] mentioned that (E) decreases with increasing the DMs content. Vafaei *et al.* (2021) [30] reported a non-significant change in the (E) modulus. One can notice that the findings are in extreme contradiction; this can be explained by the variety of the used DMs, their physio-chemical properties, and the mix design of the cementitious matrix itself.

3.2.3 Effects of DMs on Porosity and Permeability

Those two properties are of high importance since they are directly linked to the mechanical properties and to the durability performance. Some studies reported that maximizing the DMs content led to a reduction in the matrix density [9, 11, 21], where other studies have reported similar physical

TABLE 3.2 Effect of DMs on the Splitting Tensile and Flexural Strength

REF.	REPLACEMENT RATE	MIX	EFFECT ON SPLITTING TENSILE AND FLEXURAL STRENGTH
[13–15]	0, 15, 25, 35, and 50%	Mortar and concrete	The flexural and splitting tensile strength of concretes based on DMs were comparable to those for the control concrete.
[18]	0 and 20%	Mortar	The flexural strength of concrete based on DMs was higher than that of the control mixture at all curing ages.
[21]	0, 25, 50, 75, and 100%	Concrete	Tensile splitting strength tended to decrease with the increase of DMs content and water-to-cement ratio (w/c).
[22]	0 and 20%	Mortar	Flexural strength decreased with increasing the DMs content.
[23]	12.5 and 20%	Concrete	Splitting tensile strength decreased with increasing DMs content.
[7]	30%	Mortar	The mortars reached a splitting tensile of ~8 MPa at 90-d, no control was used.
[30]	100%	Concrete	Comparable flexural strength was reported.

properties [13]. These results are mainly associated with the low density of DMs compared to sand, which means higher porosity in the matrix. Table 3.3 summarizes the effect of DMs on porosity and permeability. Results show that increasing DMs content raises the total porosity. In fact, this can be explained by the higher water absorption of the DMs which lead to higher w/b ratio. The additional water dedicated to those materials' absorption evaporates with time and creates capillary porosity [33–36].

3.2.4 Effects of DMs on Durability Properties

There have been a few studies that have reported the effect of DMs as sand on the durability properties. For water absorption of hardened concrete, a similar absorption capacity of mortars and concrete based on DMs (up to 50%) compared with that of the controls was reported [13, 15]. Ozer-Erdogan et al. (2016) [21] reported that increasing DMs content leads to an increase in water absorption. Similar conclusions were brought by Safhi et al. (2022) [31].

TABLE 3.3 Effect of DM on Porosity and Permeability on the Mortars and Concrete Matrix

REF.	REPLACEMENT LEVEL	MIX	POROSITY AND PERMEABILITY
[9, 11]	0, 33, 66, and 100%	Mortar	The intrinsic permeability and the porosity increased proportionally with the increase in the content of DMs.
[13, 15]	0, 15, 25, 35, and 50%	Concrete	The use of DMs reduced the accessible pores, the sorptivity, and the water penetration depth under pressure.
[19]	0 and 100%	Mortar	The total porosity of mortars formulated with total DMs was higher than the control mortars.
[21]	0, 25, 50, 75, and 100%	Concrete	The higher w/c ratio, the higher water permeability and permeable voids volume when increasing DM content.
[23]	12.5 and 20%	Concrete	Increasing the content of DMs increased the total porosity.
[29]	100%	Concrete	DMs-based concretes show similar or slightly higher porosity.

Concerning the shrinkage, Agostini et al. (2007) [9] reported that introducing DMs greatly increased the drying shrinkage. At 110-d of curing age, the 100% substituted mortar was deformed up to nine times greater than that of the control mortar. Beddaa et al. (2020–2021) [27, 29] also reported an increase in shrinkage, which was linked to the higher absorption rate especially with the rich-fine DMs. Only one study has reported the effect of DMs as fine aggregates on the resistance to acid, which showed that the addition of 20% DMs increased the mortar's resistance to the acid attack [17]. A better resistance was observed with an increase of 28% and 42% compared to the control at 118-d for a HCl and H_2SO_4 attack, respectively. Resistance to freeze-thaw attack was studied by Junakova and Junak (2015) [18] on blended concrete samples (up to 20% DMs). The results showed that all the tested concretes reached frost resistance coefficient values > 0.85 in accordance with the standard requirements. Achour et al. (2019) [23] reported that increasing the DMs' content up to 20% substitution rate decreases the resistance to the freeze-thaw attack including the evolution of (E) modulus. Safhi et al. (2022) [31] also noticed a decrease in the dynamic elastic modulus, with up to 14% mass loss after 160 cycles. Concerning the resistance to external sulfate attack, it was found that maximizing the sediment content decreases this resistance [23].

3.3 VALORIZATION OF DREDGED MATERIALS AS ARTIFICIAL LIGHTWEIGHT AGGREGATES (LWA)

LWA can be used as a resource to produce concrete and bricks. These materials are lighter and more porous compared to the dense aggregates i.e., sand, gravel, etc. In general, two kinds of LWA can be defined, i.e., natural and synthetic. Natural LWA are composed of natural rocks including pumice, scoria, etc. Synthetic LWA are typically manufactured by expanding raw materials from thermal environments such as shales, clays, and slates. The most important process in the production of LWA is thermal expansion.

3.3.1 Production and Properties of Artificial LWA

Generally, the process technology behind this production encompasses dewatering until the moisture reaches the limit of fluidity with pelletizing or extrusion using an extruder under a precise pressure. This was in conjunction with thermal treatment (1000–1200 °C) in a rotary kiln to achieve decontaminated DMs [37–45]. Azrar et al. (2016) [46] cured pellets of DMs for 24 hours in a maintained oven at 40 °C. In other studies, DMs have been mixed with a hydraulic binder and cured in an ambient temperature of 20 °C for up to 90-d [47, 48]. In a study by Ennahal et al. (2020) [49], when DMs were mixed with thermoplastic wastes, a vacuum extrusion at 200 °C was performed. The production process of the LWA influences their physical properties.

Only 18 papers were found on this topic. The first paper found on recycling DMs as raw materials to produce LWA was published by Collins [45] in 1980. The author manufactured LWA by extrusion from dredged silt from several locations in Britain. At that time, the study encouraged this remediation pathway, especially for countries with a shortage of aggregates. The second publication on the same topic was released by Derman and Schlieper in 1999 [41] in the context of US harbors. Starting from 2006, studies started to increase at a rate of one paper per year. Figure 3.2 represents the visual aspect of LWA based on DMs with a maximum size of 12.7 mm [39]. SEM micrographs show high porosity in those aggregates. Table 3.4 summarizes the average physical properties of synthesized LWA. The studies reported a wide range in the specific gravity of 0.88–2.51, a bulk density of 544–1500 kg/m^3, and a water absorption of 0.6–51%.

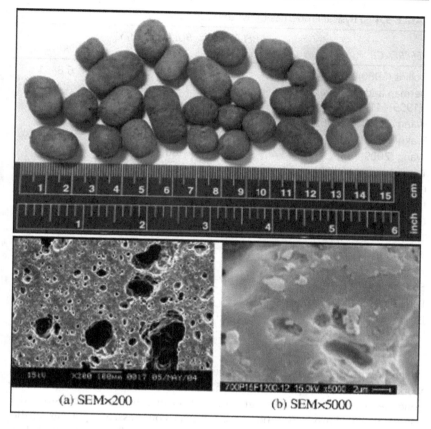

(a) SEM×200   (b) SEM×5000

FIGURE 3.2   SEM micrographs and the appearance of sintered sedimentary LWA [39]. Reprinted from Cem. Concr. Compos. 33, Tang C.-W. *et al.* "Production of synthetic lightweight aggregate using reservoir sediments for concrete and masonry", Copyright 2011.

3.3.2 Effects of LWA on the Matrix Properties

The low density with the high-water absorption of LWA have a great impact on the concrete based on those aggregates. For the microstructure, increasing the heating temperature tends to reduce the bulk density with a sharp growth in water sorption capability [37]. The high absorption of LWA requires increasing the w/b ratio thus decreasing the density and lowering the ultrasonic pulse velocity [21–23]. This absorption rate influences the fresh properties of the concrete. Hwang *et al.* (2012) [42] formulated a self-consolidating lightweight

TABLE 3.4   Physical Properties of Synthesized LWA

REFERENCE	RELATIVE GRAVITY	BULK DENSITY, kg/m³	WATER ABSORPTION, %
Collins (1980) [45]	1.55	930–1610	6.8–51
Derman and Schlieper (1999) [41]	1.34	544–880	13
Wang and Tsai (2006) [50]	–	800–1500	4.2–9.6
Brakni et al. (2009) [47]	2.48–2.51	1300–1400	40–50
Wang (2009) [38]	–	800–1100	6.4–8.9
Tang et al. (2011) [39]	–	1000–1430	10–12
Chen et al. (2012) [40]	1.00–1.41	622–859	5.5–9.5
Hwang et al. (2012) [42]	0.88–1.69	–	8.4–26
Chiou and Chen (2013) [43]	1.65–2.56	–	0.6–3.0
Tuan et al. (2014) [44]	1.25	776	2.1
Azrar et al. (2016) [46]	–	–	35
Peng et al. (2017) [48]	–	795–874	8.4–24
Ennahal et al. (2020) [49]	1.24–1.30	898–932	5.7–7.2
Peng et al. (2020) [51]	–	1010	9.4
Lim et al. (2020) [52]	1.72–2.07	–	2.2–17
Wan et al. (2022) [53]	–	1009–1098	5.4–6.4

concrete with 50, 70, and 90% replacement level. Concretes had an optimal workability. A unit weight of 1878–2057 kg/m³ resulted in high-slump flowing concrete that did not bleed or segregate. Concerning the mechanical strengths, concretes with lower w/b ratios have a higher compressive, splitting tensile, and flexural strengths than other concretes [38–40, 42]. Peng et al. (2017) [48] studied the effect of different granulometries on the mechanical properties and found that LWA with d_{max}< 3 mm resulted in better performances compared with other concretes. Using developed LWA of a density of 800 kg/m³, Wang and Tsai (2006) [50] succeeded in reaching a 28-d compressive strength of 20–30 MPa, while reaching 30–40 MPa with the LWA with a higher density of 1100–1500 kg/m³. Lim et al. (2020) [52] reported a strong correlation between open porosity and water absorption (R^2=0.99), and between compressive strength and shrinkage (R^2=0.89). Some studies reported few durability indicators, i.e., concretes with lower w/b ratio had higher electrical resistivity [38–40]. The concrete produced by Hwang et al. (2012) [42] showed good corrosion resistance which qualifies it to be classified as good quality with an electrical resistivity of> 8.5 kΩ.cm. The developed LWA concrete by Wang and Tsai (2006) [50] reached an electrical resistivity > 40 kΩ.cm after 90-d of curing. Moreover, LWA with reduced w/b ratio to 0.28 has the fewest pores for penetration and thus the least chloride penetration [37].

3.4 CONCLUSIONS

The findings on compressive, flexural, splitting tensile strengths suggested a substitution rate of fine aggregates by DMs up to 20–50% without significantly affecting the mechanical performance. Elastic properties (i.e., Young's modulus) have not shown to be significantly affected at these substitution rates. However, by increasing DMs content, the literature has revealed a decrease in density and an increase in total porosity. Also, DMs have been found to have a negative effect on durability properties such as water absorption, shrinkage, resistance to acid, freeze-thaw attack, and external sulfate attack. The main issue with using produced LWA based on DMs is their higher water absorption capacity, which decreases the density, the UPV, the mechanical strength and the electrical resistivity, and increases the porosity and the penetrability of chlorides. It can be noticed that such valorization of DMs as filler, fine, and lightweight aggregates is not often investigated. Based on this, in-depth research is needed for large-scale application and long-term durability (of concrete, bricks, etc.) exposed to extreme environmental conditions.

CREDIT AUTHORSHIP CONTRIBUTION STATEMENT

Safhi A.: Conceptualization, investigation, methodology, data curation, visualization, writing—original draft. **Ez-Zaki H.:** Validation, visualization, writing—original draft. **Rivard P.:** Validation, writing—review and editing. **Benzaazoua M.:** Validation, writing—review and editing.

REFERENCES

1. F. Rozas, A. Castillo, I. Martínez, M. Castellote, Guidelines for assessing the valorization of a waste into cementitious material: Dredged sediment for production of self compacting concrete, *Mater. Constr.* 65 (2015) 057. https://doi.org/10.3989/mc.2015.10613.
2. E. Rozière, M. Samara, A. Loukili, D. Damidot, Valorisation of sediments in self-consolidating concrete: Mix-design and microstructure, *Constr. Build. Mater.* 81 (2015) 1–10. https://doi.org/10.1016/j.conbuildmat.2015.01.080.

3. A. Dixit, H. Du, S.D. Pang, Marine clay in ultra-high performance concrete for filler substitution, *Constr. Build. Mater.* 263 (2020) 120250. https://doi.org/10.1016/j.conbuildmat.2020.120250.

4. H. Li, F. Huang, Y. Xie, Z. Yi, Z. Wang, Effect of water–powder ratio on shear thickening response of SCC, *Constr. Build. Mater.* 131 (2017) 585–591. https://doi.org/10.1016/j.conbuildmat.2016.11.061.

5. N.P. Ouédraogo, F. Becquart, M. Benzerzour, N.-E. Abriak, Influence of fine sediments on rheology properties of self-compacting concretes, *Powder Technol.* 392 (2021) 544–557. https://doi.org/10.1016/j.powtec.2021.07.035.

6. H. Beddaa, A. Ben Fraj, F. Lavergne, J.-M. Torrenti, Reuse of untreated fine sediments as filler: Is it more beneficial than incorporating them as sand?, *Buildings.* 12 (2022) 211. https://doi.org/10.3390/buildings12020211.

7. N.P. Ouédraogo, F. Becquart, M. Benzerzour, N.-E. Abriak, Environmental assessment, mechanical behavior, and chemical properties of self-compacting mortars (SCMs) with harbor dredged sediments to be used in construction, *Environ. Sci. Pollut. Res.* (2021). https://doi.org/10.1007/s11356-020-12279-6.

8. G. Xu, Z. Feng, J. Yin, W. Han, S. Ahmed, Y. Miao, Effect of salinity on rheological behavior of cement-treated dredged clays as fills, *J. Mater. Civ. Eng.* 32 (2020) 04020269. https://doi.org/10.1061/(ASCE)MT.1943-5533.0003376.

9. F. Agostini, F. Skoczylas, Z. Lafhaj, About a possible valorisation in cementitious materials of polluted sediments after treatment, *Cem. Concr. Compos.* 29 (2007) 270–278. https://doi.org/10.1016/j.cemconcomp.2006.11.012.

10. V. Dubois, R. Zentar, N.-E. Abriak, P. Grégoire, Fine sediments as a granular source for civil engineering, *Eur. J. Environ. Civ. Eng.* 15 (2011) 137–166. https://doi.org/10.1080/19648189.2011.9693315.

11. F. Agostini, C.A. Davy, F. Skoczylas, Th. Dubois, Effect of microstructure and curing conditions upon the performance of a mortar added with Treated Sediment Aggregates (TSA), *Cem. Concr. Res.* 40 (2010) 1609–1619. https://doi.org/10.1016/j.cemconres.2010.07.003.

12. H. Oh, J. Lee, N. Banthia, S. Talukdar, An experimental study of the physicochemical properties of a cement matrix containing dredged materials, *Mater. Sci. Appl.* 2 (2011) 847–857. https://doi.org/10.4236/msa.2011.27115.

13. J. Limeira, M. Etxeberria, L. Agulló, D. Molina, Mechanical and durability properties of concrete made with dredged marine sand, *Constr. Build. Mater.* 25 (2011) 4165–4174. https://doi.org/10.1016/j.conbuildmat.2011.04.053.

14. J. Limeir, L. Agulló, M. Etxeberria, Dredged marine sand as construction material, *Eur. J. Environ. Civ. Eng.* 16 (2012) 906–918. https://doi.org/10.1080/19648189.2012.676376.

15. J. Limeira, L. Agullo, M. Etxeberria, Dredged marine sand in concrete: An experimental section of a harbor pavement, *Constr. Build. Mater.* 24 (2010) 863–870. https://doi.org/10.1016/j.conbuildmat.2009.12.011.

16. Z. Lafhaj, Z. Duan, I. Bel Hadj Ali, G. Depelsenaire, Valorization of treated river sediments in self compacting materials, *Waste Biomass Valorization.* 3 (2012) 239–247. https://doi.org/10.1007/s12649-012-9110-1.

17. F. Kazi Aoual-Benslafa, D. Kerdal, B. Mekerta, A. Semcha, The use of dredged sediments as sand in the mortars for tunnel lining and for environmental protection, *Arab. J. Sci. Eng.* 39 (2014) 2483–2493. https://doi.org/10.1007/s13369-013-0805-9.

18. N. Junakova, J. Junak, M. Balintova, Reservoir sediment as a secondary raw material in concrete production, *Clean Technol. Environ. Policy.* 17 (2015) 1161–1169. https://doi.org/10.1007/s10098-015-0943-8.

19. J. Couvidat, M. Benzaazoua, V. Chatain, A. Bouamrane, H. Bouzahzah, Feasibility of the reuse of total and processed contaminated marine sediments as fine aggregates in cemented mortars, *Constr. Build. Mater.* 112 (2016) 892–902. https://doi.org/10.1016/j.conbuildmat.2016.02.186.

20. N. Manap, S. Polis, K. Sandirasegaran, M.A.N. Masrom, G.K. Chen, M.Y. Yahya, Strength of concrete made from dredged sediments. *Jurnal Teknologi* 78(7–3) (2016) 111–116. DOI: https://doi.org/10.11113/JT.V78.9496.

21. P. Ozer-Erdogan, H.M. Basar, I. Erden, L. Tolun, Beneficial use of marine dredged materials as a fine aggregate in ready-mixed concrete: Turkey example, *Constr. Build. Mater.* 124 (2016) 690–704. https://doi.org/10.1016/j.conbuildmat.2016.07.144.

22. N. Frar, H. Belmokhtar, M. Ayadi, L. Ammari, B. Allal, Valorization of port dredged sediments in cement mortars, *J. Mater. Environ. Sci.* 8 (2017) 3347–3352.

23. R. Achour, R. Zentar, N.-E. Abriak, P. Rivard, P. Gregoire, Durability study of concrete incorporating dredged sediments, *Case Stud. Constr. Mater.* 11 (2019) e00244. https://doi.org/10.1016/j.cscm.2019.e00244.

24. M. Guo, B. Hu, F. Xing, X. Zhou, M. Sun, L. Sui, Y. Zhou, Characterization of the mechanical properties of eco-friendly concrete made with untreated sea sand and seawater based on statistical analysis, *Constr. Build. Mater.* 234 (2020) 117339. https://doi.org/10.1016/j.conbuildmat.2019.117339.

25. M. Hassoune, G. Chraibi, H. Fatmaoui, J. Chaoufi, Requalification of dredging sediments through their use in concrete formulations based on a 3rd generation admixture, *Mater. Today Proc.* 22 (2020) 28–31. https://doi.org/10.1016/j.matpr.2019.08.065.

26. M. Hassoune, G. Chraibi, H. Fatmaoui, J. Chaoufi, Stability of quay wall made on concrete blocks with a formulation based on dredging sand, *Mater. Today Proc.* 36 (2021) 47–53. https://doi.org/10.1016/j.matpr.2020.05.163.

27. H. Beddaa, I. Ouazi, A. Ben Fraj, F. Lavergne, J.-M. Torrenti, Reuse potential of dredged river sediments in concrete: Effect of sediment variability, *J. Clean. Prod.* 265 (2020) 121665. https://doi.org/10.1016/j.jclepro.2020.121665.

28. F. Kazi Aoual-Benslafa, K. Touhami, Mechanical characteristics of tunnel concrete lining made with dredged sediment subjected to high temperatures, in: *Recent Advances in Environmental Science from the Euro-Mediterranean and Surrounding Regions.* 2nd Ed., Springer International Publishing, Cham, 2021: pp. 341–347. https://doi.org/10.1007/978-3-030-51210-1_56.

29. H. Beddaa, A.B. Fraj, S. Ducléroir, Experimental study on river sediment incorporation in concrete as a full aggregate replacement: Technical feasibility and economic viability, *Constr. Build. Mater.* 313 (2021) 125425. https://doi.org/10.1016/j.conbuildmat.2021.125425.

30. D. Vafaei, R. Hassanli, X. Ma, J. Duan, Y. Zhuge, Sorptivity and mechanical properties of fiber-reinforced concrete made with seawater and dredged sea-sand, *Constr. Build. Mater.* 270 (2021) 121436. https://doi.org/10.1016/j.conbuildmat.2020.121436.

31. AEM. Safhi, P. Rivard, M. Benzerzour, N.-E. Abriak, Artificial rocks based on dredged sands from the Magdalen islands in Canada: Preliminary study, Proceedings of the International Conference on Advances in Sustainable Construction Materials and Structures, Mérida, Mexico, (2021).

32. B. Vakhshouri, S. Nejadi, Review on the mixture design and mechanical properties of the lightweight concrete containing expanded polystyrene beads, *Aust. J. Struct. Eng.* 19 (2018) 1–23. https://doi.org/10.1080/13287982.2017.1353330.

33. el M. Safhi, P. Rivard, A. Yahia, M. Benzerzour, K.H. Khayat, Valorization of dredged sediments in self-consolidating concrete: Fresh, hardened, and microstructural properties, *J. Clean. Prod.* 263 (2020) 121472. https://doi.org/10.1016/j.jclepro.2020.121472.

34. el M. Safhi, P. Rivard, A. Yahia, K. Henri Khayat, N.-E. Abriak, Durability and transport properties of SCC incorporating dredged sediments, *Constr. Build. Mater.* 288 (2021) 123116. https://doi.org/10.1016/j.conbuildmat.2021.123116.

35. A. Bouchikhi, A. el M. Safhi, P. Rivard, R. Snellings, N.-E. Abriak, Fluvial sediments as SCMs: Characterization, pozzolanic performance, and optimization of equivalent binder, *J. Mater. Civ. Eng.* 34 (2022) 04021430. https://doi.org/10.1061/(ASCE)MT.1943-5533.0004071.

36. M. Amar, M. Benzerzour, A.E.M. Safhi, N.-E. Abriak, Durability of a cementitious matrix based on treated sediments, *Case Studies in Construction Materials.* 8 (2018), 258–276. DOI: doi.org/10.1016/j.cscm.2018.01.007.

37. Y.-L. Wei, J.-C. Yang, Y.-Y. Lin, S.-Y. Chuang, H.P. Wang, Recycling of harbor sediment as lightweight aggregate, *Mar. Pollut. Bull.* 57 (2008) 867–872. https://doi.org/10.1016/j.marpolbul.2008.03.033.

38. H.-Y. Wang, Durability of self-consolidating lightweight aggregate concrete using dredged silt, *Constr. Build. Mater.* 23 (2009) 2332–2337. https://doi.org/10.1016/j.conbuildmat.2008.11.006.

39. C.-W. Tang, H.-J. Chen, S.-Y. Wang, J. Spaulding, Production of synthetic lightweight aggregate using reservoir sediments for concrete and masonry, *Cem. Concr. Compos.* 33 (2011) 292–300. https://doi.org/10.1016/j.cemconcomp.2010.10.008.

40. H.-J. Chen, M.-D. Yang, C.-W. Tang, S.-Y. Wang, Producing synthetic lightweight aggregates from reservoir sediments, *Constr. Build. Mater.* 28 (2012) 387–394. https://doi.org/10.1016/j.conbuildmat.2011.08.051.

41. J.D. Derman, H.A. Schlieper, *Decontamination and Beneficial Reuse of Dredged Material Using Existing Infrastructure for the Manufacture of Lightweight Aggregate, Journal of Dredging Engineering,* 1(2) (1999).

42. C.-L. Hwang, L.A.-T. Bui, K.-L. Lin, C.-T. Lo, Manufacture and performance of lightweight aggregate from municipal solid waste incinerator fly ash and reservoir sediment for self-consolidating lightweight concrete, *Cem. Concr. Compos.* 34 (2012) 1159–1166. https://doi.org/10.1016/j.cemconcomp.2012.07.004.

43. I.J. Chiou, C.H. Chen, Effects of waste-glass fineness on sintering of reservoir-sediment aggregates, *Constr. Build. Mater.* 38 (2013) 987–993. https://doi.org/10.1016/j.conbuildmat.2012.09.042.

44. B.L. Anh Tuan, M.G. Tesfamariam, Y.-Y. Chen, C.-L. Hwang, K.-L. Lin, M.-P. Young, Production of lightweight aggregate from sewage sludge and reservoir sediment for high-flowing concrete, *J. Constr. Eng. Manag.* 140 (2014) 04014005. https://doi.org/10.1061/(ASCE)CO.1943-7862.0000835.

45. R.J. Collins, Dredged silt as a raw material for the construction industry, *Resour. Recovery Conserv.* 4 (1980) 337–362. https://doi.org/10.1016/0304-39 67(80)90039-6.

46. H. Azrar, R. Zentar, N.-E. Abriak, The effect of granulation time of the pan granulation on the characteristics of the aggregates containing dunkirk sediments, *Procedia Eng.* 143 (2016) 10–17. https://doi.org/10.1016/j.proeng.2016.06.002.

47. S. Brakni, N.E. Abriak, A. Hequette, Formulation of artificial aggregates from dredged harbour sediments for coastline stabilization, *Environ. Technol.* 30 (2009) 849–854. https://doi.org/10.1080/09593330902990154.

48. X. Peng, Y. Zhou, R. Jia, W. Wang, Y. Wu, Preparation of non-sintered lightweight aggregates from dredged sediments and modification of their properties, *Constr. Build. Mater.* 132 (2017) 9–20. https://doi.org/10.1016/j.conbuildmat.2016.11.088.

49. I. Ennahal, W. Maherzi, M. Benzerzour, Y. Mamindy, N.-E. Abriak, Performance of lightweight aggregates comprised of sediments and thermoplastic waste, *Waste Biomass Valorization.* 12 (2020) 515–530. https://doi.org/10.1007/s12649-020-00970-1.

50. H.Y. Wang, K.C. Tsai, Engineering properties of lightweight aggregate concrete made from dredged silt, *Cem. Concr. Compos.* 28 (2006) 481–485. https://doi.org/10.1016/j.cemconcomp.2005.12.005.

51. Y. Peng, X. Peng, M. Yang, H. Shi, W. Wang, X. Tang, Y. Wu, The performances of the baking-free bricks of non-sintered wrap-shell lightweight aggregates from dredged sediments, *Constr. Build. Mater.* 238 (2020) 117587. https://doi.org/10.1016/j.conbuildmat.2019.117587.

52. Y.C. Lim, Y.-J. Shih, K.-C. Tsai, W.-D. Yang, C.-W. Chen, C.-D. Dong, Recycling dredged harbor sediment to construction materials by sintering with steel slag and waste glass: Characteristics, alkali-silica reactivity and metals stability, *J. Environ. Manage.* 270 (2020) 110869. https://doi.org/10.1016/j.jenvman.2020.110869.

53. Q. Wan, C. Ju, H. Han, M. Yang, Q. Li, X. Peng, Y. Wu, An extrusion granulation process without sintering for the preparation of aggregates from wet dredged sediment, *Powder Technol.* 396 (2022) 27–35. https://doi.org/10.1016/j.powtec.2021.10.030.

Overview on Valorization of Dredged Materials as Cementitious Resource

4

Amine El Mahdi Safhi,
Abdelhadi Bouchikhi, Hassan Ez-Zaki,
Patrice Rivard

Contents

DOI: 10.1201/9781003315551-4

ABSTRACT

The production of cementitious materials, especially clinker, releases 5–8% of anthropologic carbon dioxide (CO_2) emissions. To achieve a sustainable transition to 2 °C emissions by 2050, substitution of clinker can contribute approximately 37% to the reduction of CO_2 releases [1]. The utilization of alternative raw materials for the production of clinker or the employment of supplementary cementitious materials (SCMs) could reduce the amount of clinker produced by creating hydraulic recomposed binders. Creating this last group of binders is a strategy that is highly recommended [2, 3]. An ambitious goal of reducing the average global clinker factor from 0.78 to 0.60 has been set for 2050, based on research progress on cement substitutes [1, 4]. The behavior of sediments, especially when treated, showed a pozzolanic reactivity and a potential to be used as SCMs, a raw material to produce clinker, and as a binder in geopolymers.

4.1 DREDGED MATERIALS AS SCMS FOR BLENDED CEMENTS

The sustainability requirement can be practiced in the construction field by reducing the quantity of concrete in the structure, cement in concrete, and clinker in cement by employing SCMs. The limitations are the availability of conventional SCMs e.g., fly ash (FA), metakaolin, natural pozzolan, etc. Figure 4.1 compares the availability power of dredged materials (DMs) compared with conventional SCMs. It can be concluded that those materials could be the only internationally available raw materials that can be used for blended binders' production. The first attempt to use DMs as SCMs was conducted by Dang *et al.* (2013) [6], and since then, several studies have been conducted on such upgrades. Previous studies showed that DMs demonstrate good pozzolanic activity after a convenient stimulation: Thermal calcination. A recent

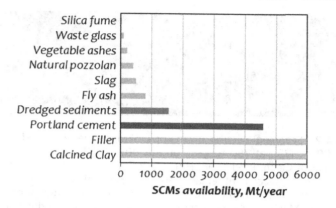

FIGURE 4.1   Estimated availability of possible SCMs. The illustrated DMs quantity is only the one deposed in the OSPAR Maritime Area [5]. Adapted from Cem. and Con. Res. (114), UN Environment *et al.*, "Eco-efficient cements: Potential economically viable solutions for a low-CO_2 cement-based materials industry". Copyright 2018.

review study on the implementation of DMs as SCMs have listed, summarized, and detailed different types of treatments. In their evaluation, several treatment techniques have been reported including physical, physio-chemical, biological, and thermal treatments [7].

4.1.1 Properties of DMs as SCMs

4.1.1.1 Morphology and Physical Characterization

DMs are very heterogeneous with as variable a morphology as their chemical composition (Figure 4.2). According to the morphological aspect, treated DMs, more specifically crushed and ground, are angular particles with irregular shape. Mostly, the specific gravity of treated DMs is below that of cement, except in one case where it was above 3.33 g/cm³ due to the high content of iron (32 g/kg) [6].

Figure 4.3 illustrates the particle size distribution (PSD) of treated DMs reported in previous sources. The 1st and 9th centiles (C_1 and C_9) were calculated to eliminate the extreme cases in the distribution. It was noticed that the DM particles under 45 μm in the range of 59–100% ($\bar{x}=82.3\%$) are in the C_1–C_9 range. This granular distribution goes along with the recommendation of ASTM C618 [8]: A good SCM should retain less than 34% on 45 μm. Table 4.1 shows the physical properties of treated DMs used as SCMs. Those

FIGURE 4.2 Scanning electron microscopy (SEM) image of treated DMs, from left: at 800 °C [25], at 750 °C [26], and at 905 °C [15]. Left figure reprinted from Journal of Clean. Prod., 198, H. Du and S. D. Pang, "Value-added utilization of marine clay as cement replacement for sustainable concrete production". Copyright 2018. Middle figure reprinted from International Journal of Concrete Structures and Materials, 44, O. Safer *et al.*, "Valorization of Dredged Sediments as a Component of Vibrated Concrete: Durability of These Concretes Against Sulfuric Acid Attack". Copyright 2018, Open Access. Right figure reprinted from Applied Clay Sci., 129, R. Snellings *et al.*, "Properties and pozzolanic reactivity of flash calcined dredging sediments". Copyright 2016.

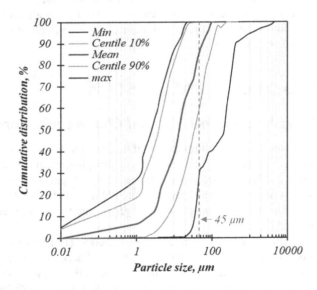

FIGURE 4.3 PSD of 25 DMs used as SCMs from 15 publications [6, 9, 11, 13, 17, 19, 21, 24–26, 29–31, 33, 38].

TABLE 4.1 Physical Properties of Treated DMs Used as SCMs

MATERIALS		DENSITY, g/cm³	SSA		d_{50}, μm
			BLAINE, cm²/g	BET, m²/g	
Dang et al. (2013) [6]	FL650	3.33	6018	–	10.3
	FL850	3.52	4275	–	13.8
Benslafa et al. (2015) [9]	DS	2.45	2990	–	20.0
Bouhamou et al. (2015–2016) [10, 11]	–	2.62	7964	–	10.1
Laoufi et al. (2016) [12]	NM	2.67	4500	–	–
	CM750	2.79	4110	–	–
	CM850	2.77	4030	–	–
	CM950	2.78	3501	–	–
Ez-zaki et al. (2016–2018) [13, 14]	DM M	–	–	–	7.36
	DM L	–	–	–	6.41
Snellings et al. (2016–2017) [15, 16]	DM	2.62	–	4.94	16.8
Benzerzour et al. (2017–2018) [17, 18]	RS120	2.48	–	8.62	2.55
	SC850T1	2.94	–	0.94	4.13
Junakova and Junak (2017) [19, 20]	DS	–	–	–	9.46
Amar et al. (2017–2018) [21–23]	RS	2.48	10093	8.62	2.53
	STDC	2.94	3996	2.06	3.69
	STFC	2.65	4106	5.99	6.36
Zhao et al. (2018) [24]	–	2.48	8200	9.10	6.58
Du and Pang (2018) [25]	–	2.87	–	–	17.8
Safer et al. (2018) [26]	–	2.65	7830	–	51.6
Van Bunderen et al. (2019–2021) [27, 28]	CFC	–	–	4.94	16.8
Safhi et al. (2018–2019) [29, 30]	TMS	2.85	–	2.34	9.80
Safhi et al. (2020–2021) [31, 32]	TMS	2.70	–	3.14	17.0
Hadj Sadok et al. (2021) [33]	–	2.63	–	–	5.05
Hadj Sadok et al. (2021) [34]	CS	2.78	6195	5.99	11.6
Kou et al. (2021) [35]	DS	1.15	–	1.70	20.3

(Continued)

TABLE 4.1 (CONTINUED)   Physical Properties of Treated DMs Used as SCMs

MATERIALS		DENSITY, g/cm³	SSA BLAINE, cm²/g	BET, m²/g	d$_{50}$, µm
Mehdizadeh et al. (2021) [36]	UFS	1.50	88.50	–	–
Ali Halassa et al. (2021) [37]	NS	2.65	–	–	–
Bouchikhi et al. (2022) [38]	RS	2.46	3429	–	33.6
	CS-450	2.72	3839	–	24.1
	CS-550	2.71	3812	–	24.6
	CS-650	2.72	3677	–	25.5
	CS-750	2.73	3257	–	29.5
	CS-850	2.74	2801	–	33.0
	CS-950	2.744	2380	–	34.4

DMs are characterized by a density of 1.15–3.52 ($\bar{x}$=2.64) g/cm³, a Blaine specific surface area (SSA) of 88–10093 ($\bar{x}$=4876) cm²/g, a BET SSA of 0.94–9.1 ($\bar{x}$=4.87) m²/g, and a d$_{50}$ of 2.53–51.6 µm ($\bar{x}$=14.5). Those parameters are linked to each other. SCMs with higher density close to the one of cement are recommended especially when the substitution is by weight. The pozzolanic reactivity is highly influenced by SSA and granulometry i.e., smaller particle size is likely associated with higher SSA and higher reactivity.

4.1.1.2 Chemical Properties

The composition of DMs varies depending on their provenance. A database of the X-ray fluorescence (XRF) analysis was developed to present the chemical composition and Loss of Ignition (LOI) of DMs from various sources (36 papers). From the chemical composition of DMs, the supreme oxides found are 35–76% ($\bar{x}$=51) of silicon dioxide (SiO_2), 0.1–26% ($\bar{x}$=11) of calcium oxide (CaO), 1.6–23% ($\bar{x}$=11) of aluminum oxide (Al_2O_3), and 0.4–13% ($\bar{x}$=5.9) of iron oxide (Fe_2O_3). As minor oxides, DMs contain 0–6.5% ($\bar{x}$=1.2) of sulfur trioxide (SO_3), 0.3–5.5% ($\bar{x}$=1.9) of magnesium oxide (MgO), 0.4–4.1% ($\bar{x}$=2.0) of potassium oxide, (K_2O), 0–2.6% ($\bar{x}$=1.2) of sodium oxide, (Na_2O), 0.1–2.3% ($\bar{x}$=1.1) of phosphorus pentoxide (P_2O_5), and 0–1.4% ($\bar{x}$=0.7) of titanium oxide (TiO_2). It is worth mentioning that some DMs contain very low concentrations of zinc oxide (ZnO) and manganese oxide (MnO). The LOI of the analyzed raw DMs was around 15–38% ($\bar{x}$=14), however, for the treated

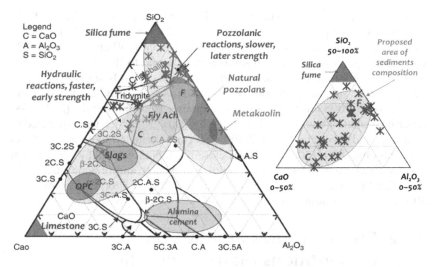

FIGURE 4.4   Composition of 39 DMs used as SCMs. Data extracted from the 32 papers adapted from [32, 39, 40].

DMs, the LOI was limited at 5.5%. The sum of percentage of pozzolanic oxides, i.e., SiO_2, Al_2O_3, and Fe_2O_3, was about 38–84% for the raw materials. For treated DMs, where the LOI is eliminated or highly reduced, this sum reached 54–96%, which classified treated DMs as a highly pozzolanic material according to ASTM C618 [8]. Figure 4.4 represents the chemical composition of the treated DMs used as SCMs in the SiO_2–Al_2O_3–Fe_2O_3 system. Regarding the ternary diagram, DMs contain 50–88% of SiO_2, 2.9–30% of Al_2O_3, and 0.1–36% of CaO.

4.1.1.3 Mineralogical Characterization

X-ray diffraction (XRD) analysis corroborate the results of XRF, i.e., DMs mainly consist of crystalline phases such as quartz and calcite, and some minor phases such as albite, anorthite, clinochlore, halite, hematite, kaolinite, muscovite, natrosilite, and nontronite. XRD analysis of calcined DMs, up to 800 °C, suggested the formation of amorphous silica as the main phase, and anhydrite, illite, and gismondine as minor phases. Only two studies have reported the Fourier-transform infrared spectroscopy (FT-IR) analysis on DMs as SCMs [33, 38]. Hadj Sadok et al. (2021) [33] reported an intense frequency due to the vibration of O–H hydroxyls and small frequencies linked to the O–H, H_2O, C–O, and Si–O bonds. The first band around 3619 cm^{-1} was attributed to O–H band and corresponds to kaolinite for the raw DMs. The vibration band at 1633 cm^{-1} was attributed to the H–O–H bond present in the raw DMs' structure. Calcite present in raw and calcined DMs was assigned to the C–O band with the vibration bands in the region of 1435–1440 cm^{-1} and 872–873 cm^{-1}. Absorption bands in both samples correspond to the elongation of the Si–O bond and the antisymmetric stretching bond of Si–O–Si in the region of 795–458 cm^{-1} indicating the presence of kaolinite. Also, the absorption band around 1000–1008 cm^{-1} corresponds to quartz. Bouchikhi *et al.* (2021) [38] reported a small vibration band around 680–700 cm^{-1}, which was attributed to kaolinite, and a vibration around 800–780 cm^{-1} attributed to quartz. A Si–O stretch around 800–1200 cm^{-1} had a close resemblance to that of 2:1 Al-rich clays such as illite and montmorillonite. Carbonates and organic matter are identified by typical vibration bands around 1420 and 2900 cm^{-1}, which are allocated to the bending vibrations of C–O and C–H, respectively.

4.1.2 Effects of DMs as SCMs on Cementitious Matrix Properties

The effect of treated DMs as SCMs has been reported including setting time, hydration heat, porosity and permeability, mechanical strength, and elastic modulus. The effect of treated DMs on durability properties has been discussed in a recent review [7], including alkali-silica reaction, accelerated carbonation, accessible porosity to water, migration of chloride ions, gas permeability, freeze-thaw test, and sulfate attacks. Also, a recent study has been conducted on the durability and transport properties of self-consolidating concrete (SCC) based on DMs [32]. For this reason, the effect of DMs as SCMs on the durability performance has not been reported in this review.

4.1.2.1 Setting Time

This test evaluates plasticity loss and is categorized by the initial setting time linked to the time of conversion from paste stage to the rigid stage, while final setting time is defined as the time from the first contact of water and cement until the paste has attained a sufficient firmness losing its plasticity against an established pressure [41]. Different parameters influencing the setting time include the fineness of the binder, the water-to-binder (w/b) ratio, and the gypsum amount in the mixture. When cement is hydrated, the gypsum reacts with C_3A quickly and generates calcium sulfoaluminate hydrate (C_4A_3S) which deposits and forms a protection layer on the cement particles to hinder the hydration of C_3A and delay the setting time of the cement. Among the 36 research studies conducted on recycling DMs as SCMs, only six have reported

TABLE 4.2   Effect of DMs on the Setting Time of Binder

REF.	REPLACEMENT LEVEL	MIX	EFFECT ON SETTING TIME
[26]	0, 10, 20, and 30%	Concrete	The setting time decreased proportionally with the increase in the fineness of cementitious materials.
[17]	0, 8, and 15%	Mortar	Mortars based on DMs showed a delay in the initial setting time (1 min per 3 g of added DMs).
[33]	0, 5, 15, and 25%	Paste and mortar	Slight effect, more precisely in the dormant period, depending on the substitution rate.
[12]	0, 10, 20, and 30%	Mortar	A proportional increase in the initial and final setting times was noticed with a rise in the substitution rate of cement by calcined mud.
[42]	0, 8, and 33%	Mortar	A delay in total setting time was linked to increasing addition amount, mainly the increase of superplasticizer associated with the addition.
[43]	0, 10, 20, 30, and 40%	Mortar	Increasing DMs content increased the initial and final setting time.

their effect on the setting time (Table 4.2). In those studies, DMs were found to generate a delay of the initial time and sometimes in the dormant period.

4.1.2.2 Hydration Heat

Table 4.3 summarizes the reported effects of DMs on the hydration heat of the blended binders, and conflicting results were drawn. Generally, the effects of DMs can be broadly classified to physical (filler role) and chemical (pozzolan role) effects. Fillers are inert materials that could improve the packing density of the particles by reducing the interstitial capillary spaces. Fillers are typically less reactive but require additional water for workability. This leads to an increase in w/b with less build-up hydrates in the mixture that cause the dilution. The degree of hydration is governed by the particle size effects and the availability of aluminates and space/pores for hydrates to precipitate. Relying on their morphology, DMs can shear the cement particles and could create additional nucleation sites. Berodier and Scrivener (2014) [44] have found that the hydration curves of cementitious paste prepared with different shearing

TABLE 4.3   Effect of DMs on the Hydration Heat of Binder

REF.	REPLACEMENT LEVEL	MIX	EFFECT ON THE HYDRATION HEAT
[25]	0 and 30%	Mortar	Pastes with 700 and 800 °C activated addition showed almost the same heat development, while paste with 600 °C activated addition possessed a lower cumulative heat.
[15]	–	–	Cumulative heat of DMs reached 300 J/g compared to 0, 100, and 1000 J/g for quartz, FA, and metakaolin respectively.
[13]	0, 8, 16, and 33%	Mortar	A significant positive effect on samples with 8 and 16% of additions, especially with the DMs treated at 850 °C.
[17]	0, 8, and 15%	Mortar	The control reached a heat of 220 J/g after 17 h 50 mins. When the mortars were substituted by 8% and 15% DMs, the heat released at the end of 15 h 36 mins and ~16 h corresponded to 262 J/g and 265 J/g, respectively.
[31]	0, 10, and 20%	SCC	The control paste generated 11.8 J/g of heat after 8.53 h, whereas pastes containing 10 and 20% of addition released 10.66 and 9.97 J/g heat after 8 h 37 mins and 8 h 34 mins, respectively.
[33]	0, 5, 15, and 25%	Paste and mortar	Released heat decreased with the increase of DMs content in the mortars.
[38]	0 and 25%	Paste and mortar	The control reached a heat flow of 7.13 J/g followed by 25% DM (750 °C) based mortar with 6.12 J/g.
[16]	0, 20, 30, and 40%	Mortar	DMs enhanced the hydration through the filler effect, and after 3–7 days contributed to the strength by pozzolanic reaction.
[28]	0, 20, 30, and 40%	Concrete	Addition of DMs and FA resulted in a lower hydration peak, due to the dilution effect.
[34]	0, 10, 15, and 20%	Mortar	The highest cumulative heat of hydration was obtained for the calcined DMs and 10% DMs mortars with 329 J/g of heat and 307 J/g of heat, while 275 J/g of heat was reported for control mortar.

(Continued)

TABLE 4.3 Effect of DMs on the Hydration Heat of Binder

REF.	REPLACEMENT LEVEL	MIX	EFFECT ON THE HYDRATION HEAT
[35]	20, 40, and 60	Pastes	All the DMs-incorporated samples suffered a decrease in heat flow and accumulated heat.
[36]	0, 5, 10, 15, 20, 25, and 30%	Mortar	Only the addition of 5% DMs slightly accelerated the rate of binder hydration. It was noticed that 10–30% of DMs acted as filler.
[42]	0, 8, and 33%	Mortar	Increased heat flow for higher replacement ratios due to the filler effect but also partly to the early pozzolanic reaction.

rates were similar to the curves in the presence of fine quartz. DMs contain 50–84% of SiO_2, however, depending on the hardness and size of the filler and the mixing process.

4.1.2.3 Porosity and Permeability

Generally, SCMs produce additional C–A–S–H or C–S–H, which improves the pore structure by densifying the cementitious matrix. Also, after the sulphates become limited, carboaluminate hydrate phases can be formed from the interaction of calcite particles with C_3A and Al from SCMs [45]. This reaction prevents the conversion of ettringite to monosulfate which can reduce the porosity of the binding phase, i.e., the density of the two types of carboaluminates (2.17 g/cm^3 for hemi-carboaluminates and 1.98 g/cm^3 for mono-carboaluminates) is higher than the density of ettringite (1.77 g/cm^3) which is beneficial for the space filling. Even so, the density of carboaluminates is still lower compared to C–S–H (2.22–2.33 g/cm^3), C–A–S–H (1.5–2.4 g/cm^3), and CH (2.2 g/cm^3) [46]. Ten studies have reported the effect of DMs on porosity and permeability (Table 4.4). Generally, increasing the incorporation rate of DMs increased the total porosity (due to the increase of the *w/b* ratio), however, the densification of the micro- and meso-pore was reported in several studies.

4.1.2.4 Mechanical Strength

Table 4.5, Table 4.6, and Table 4.7 summarize the effect of DMs on the compressive, bending, and tensile strength, respectively. Literature reported that

TABLE 4.4 Effect of DMs on the Porosity and Permeability of the Matrix

REF.	REPLACEMENT LEVEL	MIX	OPEN POROSITY, PERMEABILITY
[6]	0, 8, 16, and 33%	Mortar	Open porosity increased with the increase of DMs content (+ 17% compared to the control at 33% incorporation rate).
[13]	0, 8, 16, and 33%	Mortar	Apparent gas permeability increased moderately with the quantity of DMs. However, up to 33% of treated DMs gave lower gas permeability than the control mortar.
[17]	0, 8, and 15%	Mortar	Mortars containing DMs have higher proportion of small pores but have higher total pore area (5.53 m²/g at 28 days) compared to the control (4.57 m²/g at 28 days).
[24]	0, 10, 20, and 30%	Mortar and concrete	Total porosity increased as the substitution rate increased due to lower volume of hydrates generated by the blended binder.
[31, 32]	0, 10, and 20%	SCC	The control exhibited a lower porosity (8.74%) compared with the 10% SCC (9.08%) and 20% SCC (9.57%) samples.
[18]	0, 10, 20, and 30%	Mortar	Increasing DMs content led to an increase in porosity. 20 to 30% of DMs increased porosity by 11 and 18%, respectively.
[22]	0, 10, 20, and 30%	Mortar	Mercury porosity of the control (13.2%) was comparable to that of 10% DMs mortar (13.7%) at 60 days. Beyond this rate, the mercury porosity and water porosity increased.
[34]	0, 10, 15, and 20%	Mortar	In the long term, these substituted mortars developed a more refined pore size distribution, particularly for mortars composed of 50% cement, 40% slag, and 10% calcined DMs, and this occurs after 90-d.
[42]	0, 8, and 33%	Mortar	A comparable apparent porosity to the control mortar was achieved by the mortar containing 8% DMs.
[47]	0, 8, 16, and 33%	Mortar	Increasing the content of DMs shell powder increased the apparent porosity. The gas permeability was far lower than that of the control indicating a low porous connectivity.

TABLE 4.5   Effect of DMs on the Compressive Strength

REF.	REPLACEMENT LEVEL	TYPE OF MIX	EFFECT ON THE COMPRESSIVE STRENGTH
[6]	0, 8, 16, and 33%	Mortar	Negative effect, however, 650 °C treated DMs showed better behavior among other treated DMs.
[25]	0 and 30%	Mortar	A slight decrease was noticed (93% strength loss at 28-d).
[26]	0, 10, 20, and 30%	Concrete	Concretes with 10% addition content had lower strength at an early stage, but beyond 90-d the strength was superior to the control concrete.
[9]	0, 10, 15, and 20%	Mortar	Increasing DMs content led to up to 30% strength loss for a substitution rate of 20% at 90-d.
[10, 11]	0, 10, 15, and 20%	SCC	Increased with increasing DM content. Up to 10% substitution rate, equivalent strength was reached.
[13]	0, 8, 16, and 33%	Mortar	Decreased with increased rate of substitution. But 16% addition can be used for 42.5 class cement.
[17]	0, 8, and 15%	Mortar	DMs enhanced the resistance: At 60-d, the 8 and 15% DMs mortars reached 65 and 63 MPa respectively, where the control reached 64 MPa compressive strength.
[19]	0 and 40%	Concrete	A very negative effect, i.e., 40% substitution rate led to a significant decrease of compressive strength.
[24]	0, 10, 20, and 30%	Mortar and concrete	Strength decreased when the substitution rate increased. The compressive strength of 10% DMs mortar was slightly higher than the control mortar.
[29]	0 to 50%	Self-consolidating mortars	The variation of DMs-to-cement ratio from 0 to 0.5 is associated with a compressive strength variation ranging from 80 to 60 MPa.
[30]	0 to 30%	SCC pastes	Treated DMs generated low reduction in compressive strength compared to the w/b ratio.

(Continued)

TABLE 4.5 (CONTINUED) Effect of DMs on the Compressive Strength

REF.	REPLACEMENT LEVEL	TYPE OF MIX	EFFECT ON THE COMPRESSIVE STRENGTH
[31, 32]	0, 10, and 20%	SCC	Compressive strength of the mixture made with 10% DMs was higher than that of the control. 20% replacement led to a lower early stage strength but comparable at 90-d.
[33]	0, 5, 15, and 25%	Paste and mortar	Higher development of strength due to the cure at high temperature, which accelerated the pozzolanic activity.
[38]	0 to 30%	Mortar	At up to 20% DMs incorporation, comparable strength was found.
[12]	0, 10, 20, and 30%	Mortar	The control mortar had the highest compressive strength; however, mortars with calcined DMs at 750 °C had better performances that those obtained at 850 and 950 °C/3 h.
[16]	0, 20, 30, and 40%	Mortar	Increasing the cement replacement level from 20 to over 30–40% resulted in a similar but slower strength development up to 182 days strength compared to control.
[18]	0, 10, 20, and 30%	Mortar	Strength of mortars with 10–20% calcined DMs was comparable to that of control. At 30% DM content, a considerable drop in resistance (20%) was noticed.
[20]	0 and 40%	Concrete	Compressive strength of the mixtures with DMs composite was lower than the control nearly by half.
[23]	0, 10, 20, and 30%	Mortar	10% DMs mortar and mortar control presented a comparable compressive strength. Beyond this rate, a dramatic strength decrease was observed.
[28]	0, 20, 30, and 40%	Concrete	Slow strength development at an early stage, but higher compressive strength results are obtained at later stages.

(Continued)

TABLE 4.5 (CONTINUED) Effect of DMs on the Compressive Strength

REF.	REPLACEMENT LEVEL	TYPE OF MIX	EFFECT ON THE COMPRESSIVE STRENGTH
[34]	0, 10, 15, and 20%	Mortar	Higher strengths were obtained from mortars made of 50% OPC, 40% slag, and 10% calcined DMs, occurring after 90-d.
[35]	20, 40, and 60	Mortar	All DMs-incorporated mortars suffered a decrease in strength at all curing stages, indicating that DMs possessed a negligible pozzolanic activity.
[36]	0, 5, 10, 15, 20, 25 and 30%	Mortar	Strength decreased by adding DMs; however, the mixtures with 5–15% DMs achieved > 85% of the strength of the control after 91-d.
[42]	0, 8, and 33%	Mortar	Mortars containing 8% of treated minerals had higher compressive strength than 52 MPa, which keeps these mixtures in the same class of the ordinary mortar.
[43]	0, 10, 20, 30, and 40%	Mortar	Proportional decrease of compressive strength when increasing DMs content up to 70% for a substitution rate of 40%.
[48]	0, 10, 30, 50, and 70%	Mortar	At 28-d, 10% substitution rate exhibited higher strength than the control. Up to 30%, comparable strengths were achieved. Above 50%, the strength significantly decreased.

increasing DM content is associated with a decrease in mechanical strength. However, a substitution rate of 10–20% was found to yield a performance comparable to the controls. For higher rates, the strengths decrease dramatically.

4.1.2.5 Young's Modulus

The determination of Young's modulus helps to understand how the presence of DMs influences the behavior of the cementitious matrix under mechanical stresses (compressive or flexion strength) before the elastic deformation. It also provides information on the ability of DMs to influence the nature of the interconnections between cementitious particles and hydrate phases. Only seven

TABLE 4.6 Effect of DMs on the Flexural Strength

REF.	REPLACEMENT LEVEL	MIX	EFFECT ON THE FLEXURAL STRENGTH
[19]	0 and 40%	Concrete	Blended mixtures showed flexural strength of 14–66% lower than the control mixture.
[24]	0, 10, 20, and 30%	Mortar and concrete	Comparable flexural strength up to 10% substitution rate. At more than 10%, a dramatic decrease was noticed.
[29]	0 to 50%	Self-consolidating mortars	The DMs-to-cement ratio had a low impact on bending strength: Its variation from 0 to 0.5 yielded a bending strength variation from 10 to 11 MPa.
[12]	0, 10, 20, and 30%	Mortar	It can be stated that in general the incorporation of calcined mud helped in improving the flexural strengths, at all substitution rates.
[18]	0, 10, 20, and 30%	Mortar	10 to 20% DMs-based mortars were comparable to the control strength. There was a drop in the resistance at 30%.
[20]	0 and 40%	Concrete	Comparable flexural strength to that of the control was achieved considering the error bars.
[22]	0, 10, 20, and 30%	Mortar	Higher DMs content yielded lower strength. However, a series containing 10% DMs had substantially similar bending strengths to one of the controls.
[34]	0, 10, 15, and 20%	Mortar	It was generally expressed that the incorporation of slag and calcined DMs had limited effects on the strength after more than 7-d at all substitution rates.

(Continued)

TABLE 4.6   Effect of DMs on the Flexural Strength

REF.	REPLACEMENT LEVEL	MIX	EFFECT ON THE FLEXURAL STRENGTH
[35]	20, 40, and 60	Mortar	All DMs-incorporated mortars suffered a decrease in strength at all curing stages, indicating that DMs possessed a negligible pozzolanic activity.
[36]	0, 5, 10, 15, 20, 25 and 30%	Mortar	A low replacement ratio (5–10%) resulted in a slight enhancement in the strength at 28-d. However, at high content of DMs and longer times (> 56-d), the strength was reduced to below that of the control.
[43]	0, 10, 20, 30, and 40%	Mortar	Proportional decrease of the strength: 10% DMs incorporation resulted in 90% of compressive strength of the control.
[48]	0, 10, 30, 50, and 70%	Mortars	10% DMs mortar has higher strength than that of control. Up to 30%, the strength was comparable to that of control. Above 50%, the strength significantly decreased.

studies have reported the effect of DMs on Young's modulus (E) (Table 4.8). Generally, a slight decrease is noted in the major studies (compared to matrix of control: 100% cement), and this observation remains compatible with the nature of the DMs, which require time to start to react with the components of the matrix (by pozzolanic effect).

4.1.3 Assessing the Recycling of DMs

The Figure 4.5 shows a flowchart proposed to assess the feasibility of recycling DMs as SCMs. The first step is the characterization of the raw material i.e., the physical, chemical, mineralogical, and environmental aspects. At least 90% of the particles should be smaller than 80 μm with a sum of Al_2O_3, SiO_2, and Fe_2O_3 oxides higher than 50%. At this step, even though heavy metals and

TABLE 4.7 Effect of DMs on the Tensile Strength

REF.	REPLACEMENT LEVEL	MIX	EFFECT ON THE TENSILE STRENGTH
[10, 11]	0, 10, 15, and 20%	SCC	Strength decreased with increasing DMs content. Up to 10% substitution rate, strengths were still comparable to the control.
[31, 32]	0, 10, and 20%	SCC	Similar behavior up to 10% substitution by DMs. At 20% replacement, there was a slight decrease of strength at an early stage but still acceptable considering the standard deviation.
[28]	0, 20, 30, and 40%	Concrete	The control had the highest tensile strength, but considering the error bars, 20% concrete has an equivalent tensile strength.

organic concentrations exceed the recommended limits, the materials could be considered. Thermogravimetric analysis coupled with differential scanning calorimetry (TGA-DSC) are required to define the suitable calcination temperature for the activation of the mineral phases and allow insights to be given about the mineralogical compounds after thermal treatment. The calcination domain is the end of decarbonation, which is considered between the end of dihydroxylation and the beginning of recrystallization.

Once calcination is complete, grinding might be required, not only to prevent the agglomeration of particles but also to target the d_{50} less than 45 μm and an SSA BET higher than 2220 m²/kg. After calcination, the sum of Al_2O_3, SiO_2, and Fe_2O_3 should not change significantly; however, according to ASTM C618, the concentration of SO_3 and alkali as Na_2O should not exceed 5% and 1.5%, respectively [8]. Once again, heavy metal concentration should be determined to quantify the mobility of contamination or toxicity of DMs. Even though the material is not inert but also not dangerous, it could be considered depending on the incorporation rate. In this case, final compliance testing must be carried out on the final product. However, if the material is found to be hazardous [49, 50], it should be discarded unless decontamination is considered to be feasible.

The pozzolanic reactivity of calcined DMs is very important for them to be considered as SCMs. This potential could be tested using direct methods which mainly focus on CH consumption in hydrated paste (XRD, TGA, le Chapelle test, Frattini test, and scanning electron microscopy). Indirect methods can also be tested, including lime reactivity, strength activity index, electric conductivity, calorimetry, flocculation test, and the R^3 test method [51]. If this potential is found to be comparable to the cement, calcined DMs could be used as SCMs. If not, they could be considered as a mineral addition or filler.

TABLE 4.8   Effect of DMs on Young's Modulus

REF.	REPLACEMENT LEVEL	MIX	EFFECT ON YOUNG'S MODULUS
[6]	0, 8, 16, and 33%	Mortar	DMs decreased the elastic modulus proportionally with the increase of the replacement level.
[17]	0, 8, and 15%	Mortar	At 28-d, 8% DMs-based mortar reached 42 GPa, while the control reached 44 GPa. The measured (E) for 15% DMs-based mortars was the lowest.
[29]	0 to 50%	Self-consolidating mortars	Decreased from 56 to 45 GPa at 90 days, for DMs-to-cement ratio of 0 to 0.5, respectively.
[31]	0, 10, and 20%	SCC	10 to 20% DMs-based SCCs have slow development of (E) but reached the control starting from the 56th day.
[22]	0, 10, 20, and 30%	Mortar	At 90 days, 10% DMs-based mortars and the control showed similar (E) of 44 GPa.
[28]	0, 20, 30, and 40%	Concrete	The static E-modulus showed a faster development, and less differences were observed between the different mixes.
[34]	0, 10, 15, and 20%	Mortar	After 28-d, all mortar with slag and calcined DMs seemed to show the same values of dynamic Young's modulus.

The material should be tested at final product level. Workability, mechanical performance, and durability properties should be tested for compliance with technical specifications. If the technical testing is satisfactory, then an environmental inspection follows. If the heavy metals are not stabilized in the matrix, the material can be calcined or pretreated differently.

4.2 VALORIZATION AS RAW MATERIAL FOR CLINKER PRODUCTION

Dredged sediments typically have a mineral composition rich in SiO_2, Al_2O_3, Fe_2O_3, and CaO, which are essential to cement manufacture. Considering

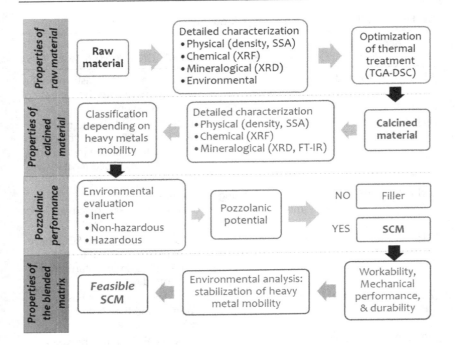

FIGURE 4.5 Feasibility of recycling DMs in a cementitious matrix.

the DMs' composition and other available raw materials, the proportion of all other ingredients can be adjusted to maximize the incorporation of DMs. Some relations were found between the percentage of those oxides, including lime saturation factor (LSF), silica ratio (SR), and alumina ratio (AR). Other factors are less considered, such as hydraulic modulus (HM), liquid phase, and degree of sulfurization.

$$LSF = \frac{CaO}{2.8SiO_2 + 1.18Al_2O_3 + 0.65Fe_2O_3} \tag{4.1}$$

$$SR = \frac{SiO_2}{Al_2O_3 + Fe_2O_3} \tag{4.2}$$

$$AR = \frac{Al_2O_3}{Fe_2O_3} \tag{4.3}$$

Clinkerization requires a high temperature (1450 °C) to convert raw materials to the clinker composed of alite ($3CaO \cdot SiO_2$), belite ($2CaO \cdot SiO_2$), aluminate

$(3CaO \cdot Al_2O_3)$, and ferrite $(4CaO \cdot Al_2O_3 \cdot Fe_2O_3)$, abbreviated respectively as C_3S (40–60%), C_2S (16–30%), C_3A (7–15%), and C_4AF (7–12%). Only seven papers have reported the use of DMs as a raw material for the production of clinker [52–58]. The first paper dates back to 1999 conducted by Rehmat et al. (1999) [54], discussing the feasibility of using the so-called Cement-Lock™ Technology to produce cement from contaminated DMs. The process is based on clinkering DMs in a reactive melter at 1200–1400 °C mixed with suitable modifiers and a controlled flue gas to capture HCl. A demonstration-scale plant with about 23000 m³ per year capacity has been constructed in New York, but no characterizations or technical results have been presented.

In 2004, Dalton et al. (2004) [55] published a paper on producing Ordinary Portland Cement (OPC) made from contaminated DMs of New York harbor. Four clinkers were manufactured with DMs: 0, 1.49, 6.63, and 12.3% fired up to 1450 °C, mixed afterwards with 6.1% gypsum, and ground to a Blaine SSA of 375 m²/kg. XRD analysis showed that alite content decreases with increasing the DMs content; more precisely, the matrix with 12.3% of DMs had about a 42% lower alite content than the control. It is worth noticing that fired material at full scale had a better result than that fired in the lab, which was attributed to the cooling process. It was proved that when cooling occurs very slowly, the alite phase will decompose in 1250–1100 °C to form a blended mixture of belite and lime (CaO) [59]. A practical replacement of 3–6% of total feedstock material (dry mass basis) was recommended considering the factors of chloride scaling and SiO_2 burning.

Aouad et al. (2012) [52] investigated the feasibility of DMs as novel supply of the raw material to produce clinker. The XRF analysis showed that cement made from DMs (39 wt.%) had a comparable chemical composition to OPC. XRD analysis corroborates these findings; the crystalline phases in OPC were found in the synthesized blended cement. In terms of hydration rate, the two pastes had the same released heat. However, the blended cement had an accelerated kinetics linked to the fineness and the gypsum amount. The results showed that the blended cement exhibited compressive strengths equivalent to those obtained with OPC at early stages and 20% higher compressive strength at 56 days. This result could be explained by the cement paste XRD patterns, which suggested that silicates were more reactive than in OPC. For each metric ton of OPC produced, 390 kg of polluted fluvial DMs can be recycled. This amount was never reached in previous studies on the substitution of cement raw materials.

Faure et al. (2017) [53] investigated the ability of dam fine-grained DMs to be used as a raw material for clinker production. The DMs used contained 44% of SiO_2, 15% of CaO, 11% of Al_2O_3, with 19% LOI. DMs were used as total replacement of quarried clay (27%). This incorporation rate was calculated with respect to a fixed LSF of 98.5 and SR of 2.4. The mineralogical composition of composed clinker using optical microscopy counting was comparable to the one with the Rietveld method: Around 71% of C_3S, 11% of C_2S, 11% C_3A, and 5.5% of C_4AF. As shown in the micrographs (Figure 4.6), the porosity was about the third of the total clinker grain volume. This could be explained by the process (granulation before clinkerization). Nevertheless, SEM observations demonstrate a high proportion of relatively small (lower than 50 μm) C_3S crystals. Small alite crystals are more reactive than larger ones.

Anger *et al.* (2019) [56] studied the feasibility of using three different fine-grained reservoir DMs as raw material for clinker production. The specific values of LSF, SR, and HM were fixed for the "raw meal" of each clinker based on the three DMs. Nine raw mixes were addressed with a range of DMs' incorporation of 20–31 wt.%. The theoretical Bogue calculation revealed that the phase's composition of those clinkers is: 34–65% C_3S, 13–36% C_2S, 9–15 C_3A, and 6–12 C_4AF. The XRD analysis observed the following proportion: C_3S: 50–65%, C_2S: 15–20%, C_3A: 5–15%, and C_4AF: 5–10%. Microstructural analyses of the produced clinkers showed that all the studied DMs could be successfully recycled into clinker production. In 2019, the same research group published another paper investigating the reuse of seven dam DMs from different French regions as clinker raw material [57]. The used dam DMs contained 35–47% SiO_2, 12–23% CaO, 7.4–14% Al_2O_3, with 15–23% LOI. Two types of design were used—the first for binary clinker meals where DMs were mixed with limestone for a fixed LSF of 97. The second mix design was a ternary clinker meal where DMs replaced a calculated percentage of clay by fixing LSF at 97 and SR at 2.4. In both, 14 composed clinkers were prepared,

FIGURE 4.6   Polished surface of clinker with DMs (left) and without (right) [53]. Reprinted from Constr. Build. Mater., 65, R.V. Silva, J. de Brito, and R.K. Dhir, "Properties and composition of recycled aggregates from construction and demolition waste suitable for concrete production", 201. Copyright 2014.

with DMs' incorporation rate ranging from 10% to 35%. The theoretical Bogue calculation revealed that the phase's composition of those composed clinkers is: 52–70% C_3S, 6–26% C_2S, 6–12% C_3A, and 8–12% C_4AF. The mechanical tests and the cumulative hydration results showed that very high DMs' replacement ratios (25–35 wt.%) are reached in the total clay substitution strategy, called binary mixes. Considering the clinker obtained with ternary meals, a 10–15 wt.% replacement level of the usual raw material by any of the DMs does not significantly impact the clinker properties and the phase polymorphism. After the addition of gypsum, CEM I 52.5 N cement can be obtained with high hydraulic reactivity and no constraint in terms of workability and shrinkage.

More recently, Chu *et al.* (2022) [58] studied the feasibility of using river DMs from north of France as raw material to produce clinker. The DMs used were mainly composed of SiO_2 (~40%), CaO (~11%), Al_2O_3 (~10%), with ~28% LOI. The XRD analyses showed the presence of quartz, calcite, brushite, and muscovite. LSF of 97, SR of 2.6, and AR of 1.45 were targeted using the solver option in Microsoft Excel. The authors chose an LSF of 97 to minimize the free lime (CaO_{free}), thus maximizing the content of C_3S in the clinker. The criteria have led to an incorporation rate of DMs up to 31 wt.%. The theoretical Bogue calculation revealed that the phase's composition of the composed clinker was 67% C_3S, 11% C_2S, 7.6% C_3A, and 10% C_4AF. A numerical modeling of cement hydration was conducted by CEMHYD3D and showed that, up to the mentioned incorporation rate, the phase assemblage and the hydration behavior were not affected. The developed cement showed similar performance to that of the control; however, a decrease in the mechanical strength was observed. The authors justified this by the formation of the M_3 polymorph of C_3S in the clinker-based DMs, which is less reactive than the M_1 polymorph.

Despite the few studies conducted on using DMs for clinker production, the reported results are encouraging to conduct in-depth research and large-scale testing.

4.3 VALORIZATION AS GEOPOLYMER AND ALKALI-ACTIVATION BINDERS

Recently, more research studies were oriented toward alkali-activation and geopolymerization technology to develop eco-binders with low carbon emissions by using mineral waste such as DMs among others (FA, slags, waste glass, etc.). These new binders are characterized by low cost, low carbon emissions, and better technical properties compared to OPC. Due to the utilization of calcined clays and other materials as sources of aluminosilicate, the DMs presented a reservoir of alternative primary raw materials rich in clay phases and other elements which can be used as alkali-activated and geopolymer materials. The studies show that different DMs are mostly rich in quartz, clay, alkaline phases, etc. The concentration of these phases varied according to the origin of sedimentation.

It is considered that the evolution of the matrix properties is related to the nature of DMs, type of treatment, and the conditions of cure. The results presented in Table 4.9 show that most DMs presented a considerable source of aluminates for alkali-activated materials and geopolymer. When the DMs are calcined, the temperature of calcination was in the range of 650–750 °C. The mechanical treatment (crushing and grinding) can be adapted for the sand DMs that contain limestone. Calcination between 450 °C and 550 °C is more appropriate for DMs rich in organic matter (OM) and organic aluminosilicate, precisely because of humic and fluoric acids which contain more hydrated chemical elements. A calcination temperature up to 850 °C does not show any advantage for various DMs; on the contrary, the crystallization product caused by calcination decreases the reactivity of the clay phases [38]. Another type of treatment is characterized by mechanochemical activation alongside physical and chemical activation. This type of activation gives more homogenization and more interaction between the activators and the particle content in the DMs.

4.4 CONCLUSIONS

It was revealed that DMs used as SCMs are mainly composed of 35–76% SiO_2, 0.1–26% CaO, 1.6–23% Al_2O_3, and 0.4–13% Fe_2O_3. The LOI of the raw DMs was about 15–38%, which decreases after treatment to lower than 5.5%. The sum of SiO_2, Al_2O_3, and Fe_2O_3, was about 38–59% for the raw DMs and reached 54–97% after the treatment, which classified the treated DMs as a highly pozzolanic material. Their reported pozzolanic reaction makes them suitable materials to be used as a cementitious addition. The high SiO_2 content in DMs enables the pozzolanic reaction, with the addition of CaO producing additional C–S–H gels. However, DMs found to have a retarding effect on the setting time can either alter or accelerate the early hydration kinetics. PSD, chemical properties, and the rate of incorporation were found to be the critical factors influencing this effect.

It was identified that the use of 10–20% treated DMs is the optimal rate to produce high-strength concrete (> 50 MPa at 28-d). Indeed, increasing the content of DMs increases the total porosity and permeability; however, it

TABLE 4.9   Effect of DMs on geopolymer mixture

REF.	USED MATERIALS	TREATMENT	HIGHLIGHTS
Ferone et al. (2013) [60]	Combination of two DMs: One rich in OM (Sed1) and the other rich in $CaCO_3$ (Sed2)	Calcination for 1/2 h at 650 °C and at 750 °C + 5M-NaOH mixed solution	The best properties were observed at 750 °C treatment for cured at 60 °C for four days. Sed1 reached higher compressive strength compared with Sed2.
Ferone et al. (2015) [61]	Two different DMs with 62–67% of clay phases content with slight difference in calcite	DMs were calcined at 400 °C and 750 °C, and treated by 4 and 7M-NaOH	Addition of NaOH solution and ground granulated blast furnace slag (GGBS) allowed the obtainment of better geopolymerization, thus better mechanical performances. Increasing the calcination temperature increased the reactivity of the DMs.
Komnitsas (2016) [62]	Comparison between additional source of waste (DMs, bricks, concrete, tiles, etc.)	Activation by KOH (up 12 M). The cure was done at 80 °C for 24 h.	The best properties were achieved by mortars formed by tiles. The mechanical properties decrease after immersion in the water and seawater (up to 50% of performances).
Slimanou et al. (2019) [63]	Effect of calcination of DMs on the mechanical properties and durability in hydrochloric acid solution.	Uncalcined and calcined DMs at 750 °C were used to substitute metakaolin up to 25 wt.%. The used activator was NaOH in sodium silicate solution.	The incorporation of calcined DMs allowed reinforcing the structure by decreasing the microstructural porosity. Mechanical properties and durability in hydrochloric acid solution was enhanced.
Bouchikhi et al. (2019) [64]	Combination of waste glass and DMs in geopolymer matrix	DMs were treated at an optimal temperature of 750 °C and used to substitute the metakaolin up to 30%	The mechanical properties of the matrix increased compared to the control matrix (100% MK). 20% substitution was found to be the optimal rate.

(Continued)

TABLE 4.9 (CONTINUED) Effect of DMs on geopolymer mixture

REF.	USED MATERIALS	TREATMENT	HIGHLIGHTS
Hosseini et al. (2021) [65]	Low-reactivity DMs of high-plasticity clay, rich in quartz and alumina, was mixed with FA	DMs mixed with FA and sodium silicate gave better coiling of particles with the chemical activator (sodium silicate)	The mixture 1:1 of DMs: FA treated by mechanochemical process promoted connectivity and geopolymerization, increased the compressive strength, and reduced the pore surface area.
Karam et al. (2021) [66]	Alkali-activated binder, based on GGBFS activation, incorporating the fine fraction of uncalcined marine DMs	Treatment with two alkali activators: Sodium hydroxide solution and mixed solutions of sodium hydroxide and sodium silicate	The DMs' incorporation did not affect the structure of formed hydrates. The substitution of slag by DMs resulted in a delay in the initiation of activation only in the short term. The DMs did not contribute to the phases development.

reduces the micropore and mesopore levels. A DMs-based matrix has comparable flexural and splitting tensile strengths, with replacement levels up to 20%. Worldwide cement consumption reached almost 4144 million tons (Mt) in 2020 [67]. The results suggest a practical use of DMs as SCMs of 10–20% (dry mass basis). This material replacement rises to 414–823 Mt per year (dry mass basis). Considering 127% as a typical in-place water content of DMs, and assuming a bulk density of in-place DMs of 1300 kg/m^3 and LOI of 15–38%, this corresponds to the use of 804–2207 Mm^3 of those materials annually which represents > 80% of total worldwide DMs (estimated at > 1000 Mm^3 [17]). The same proportions apply for use of DMs as a raw material for clinker production. One can see that those two remediation pathways could consume all the generated DMs.

For a geopolymer matrix, it is recommended to grind the material (< 80 μm) to increase the contact surface with the activator and to improve the rearrangement of the particles in the matrix, i.e., a higher packing density. Thermal treatment was recommended depending on the nature, and the chemical and mineralogical composition of DMs. For sediments containing carbonate mineral phases, such as alkaline, calcite, and others, the temperature commonly used for calcination ranged from 550–750 °C. The activation of certain phases such as $CaCO_3$ and $MgCO_3$ leads to alkaline self-activation (K^+, Na^+, etc.) but also to pozzolanic self-activation (Ca^{2+}, Mg^{2+}, etc.). The presence of mineral salts, pollutant metals, and organic residues requires a pretreatment of immobilization/degradation.

4.5 CHALLENGES AND FUTURE SCOPE

As the one can see, intensive research was conducted on upcycling DMs as a cementitious resource; nevertheless, some challenges have not been tackled yet. In-depth research is needed for large-scale applications, particularly concerning long-term durability when concrete is exposed to extreme environmental conditions. Moreover, the influence of DMs on structural reinforced concrete has not been studied. Moreover, the most common research topic regarding concrete is new generation types, such as 3D-printed concrete, ultra-high-performance concrete, fiber-reinforced concrete, SCC, etc., but the potential use of DMs in those concretes remains an untouched and unexplored field. For better understanding of the effect of DMs on mechanical properties, regardless of their role, it is required to report the stress-strain curve. Thus, behavior laws could be proposed. Also, reducing the high water demand of DMs will enhance based concrete, and increase the DMs incorporation rate.

Geopolymer is one of the research trends in construction materials and can reduce up to 85% of CO_2 emissions. Only a few studies have investigated the use of DMs in geopolymer, even though DMs are a rich source of aluminosilicate, which is the primary requirement to produce geopolymer. However, in-depth analyses were lacking, and more research should be done for a proper mix design and proper curing regime. The use of DMs for clinker production is very promising and enhances the incorporation rate of those materials. However, very few studies have been conducted in such valorization. A deeper investigation is needed in terms of the durability aspect and thermodynamic modeling.

Assessment of the life cycle of DMs as cementitious resources is very necessary, including study of the energy summary (released CO_2 from the treatment process). In addition, the effect of the curing regime proved to have a significant influence on different cementitious systems; however, no study so far has focused on this topic. The use of some SCMs is possible according to regulations and standards such as EN 15167–1:2006 GGBS for use in concrete, mortar, and grout [68], EN 13263–1+A1:2009 silica fume for concrete [69], EN 450–1:2012 FA for concrete [70], and ASTM C1866–2020 Standard Specification for Ground-Glass Pozzolan for Use in Concrete [71]. Further research and efforts are required for standardization regarding the use of DMs.

CREDIT AUTHORSHIP CONTRIBUTION STATEMENT

Safhi A.: Conceptualization, investigation, methodology, data curation, formal analysis, visualization, writing—original draft. **Bouchikhi A.**: Validation, writing—original draft. **Ez-Zaki H.**: Validation, writing—review and editing. **Rivard P.**: Validation, writing—review and editing.

REFERENCES

1. Technology roadmap, in: *Springer Reference*, Springer-Verlag, Berlin/ Heidelberg, 2011. https://doi.org/10.1007/SpringerReference_7300.
2. M. Schneider, M. Romer, M. Tschudin, H. Bolio, Sustainable cement production: Present and future, *Cem. Concr. Res.* 41 (2011) 642–650. https://doi.org/10.1016/ j.cemconres.2011.03.019.
3. S.A. Miller, V.M. John, S.A. Pacca, A. Horvath, Carbon dioxide reduction potential in the global cement industry by 2050, *Cem. Concr. Res.* 114 (2018) 115–124. https://doi.org/10.1016/j.cemconres.2017.08.026.
4. International Energy Agency, *Cement Technology Roadmap: Carbon Emissions Reductions up to 2050*, OECD, 2009. https://doi.org/10.1787/9789264088061-en.
5. K.L. Scrivener, V.M. John, E.M. Gartner, Eco-efficient cements: Potential economically viable solutions for a low-CO2 cement-based materials industry, *Cem. Concr. Res.* 114 (2018) 2–26. https://doi.org/10.1016/j.cemconres.2018.03.015.
6. T.A. Dang, S. Kamali-Bernard, W.A. Prince, Design of new blended cement based on marine dredged sediment, *Constr. Build. Mater.* 41 (2013) 602–611. https://doi.org/10.1016/j.conbuildmat.2012.11.088.
7. M. Amar, M. Benzerzour, J. Kleib, N.-E. Abriak, From dredged sediment to supplementary cementitious material: Characterization, treatment, and reuse, *Int. J. Sediment Res.* 36 (2021) 92–109. https://doi.org/10.1016/j.ijsrc.2020.06.002.
8. ASTM C618, Specification for coal fly ash and raw or calcined natural pozzolan for use in concrete, *ASTM International*, 2019. https://doi.org/10.1520/C0618-19.
9. F.K.A. Benslafa, D. Kerdal, M. Ameur, B. Mekerta, A. Semcha, Durability of mortars made with dredged sediments, *Procedia Eng.* 118 (2015) 240–250. https://doi.org/10.1016/j.proeng.2015.08.423.
10. N. Bouhamou, F. Mostefa, A. Mebrouki, The influence of dredged of natural waste on shrinkage behavior of self compacting concrete for achieving environmental sustainability, *Key Eng. Mater.* 650 (2015) 91–104. https://doi.org/10. 4028/www.scientific.net/KEM.650.91.
11. N.-E. Bouhamou, F. Mostefa, A. Mebrouki, K. Bendani, N. Belas, Influence of dredged sediment on the shrinkage behavior of self-compacting concrete, *Mater. Tehnol.* 50 (2016) 127–135. https://doi.org/10.17222/mit.2013.252.

12. L. Laoufi, Y. Senhadji, A. Benazzouk, Valorization of mud from Fergoug dam in manufacturing mortars, *Case Stud. Constr. Mater.* 5 (2016) 26–38. https://doi. org/10.1016/j.cscm.2016.06.002.

13. H. Ez-zaki, A. Diouri, S. Kamali-Bernard, O. Sassi, Composite cement mortars based on marine sediments and oyster shell powder, *Mater. Constr.* 66 (2016) e080. https://doi.org/10.3989/mc.2016.01915.

14. H. Ez-zaki, A. Diouri, Chloride penetration through cement material based on dredged sediment and shell powder, *J. Adhes. Sci. Technol.* 32 (2018) 787–800. https://doi.org/10.1080/01694243.2017.1378068.

15. R. Snellings, Ö. Cizer, L. Horckmans, P.T. Durdziński, P. Dierckx, P. Nielsen, K. Van Balen, L. Vandewalle, Properties and pozzolanic reactivity of flash calcined dredging sediments, *Appl. Clay Sci.* 129 (2016) 35–39. https://doi.org/10.1016/j. clay.2016.04.019.

16. R. Snellings, L. Horckmans, C. Van Bunderen, L. Vandewalle, Ö. Cizer, Flash-calcined dredging sediment blended cements: Effect on cement hydration and properties, *Mater. Struct.* 50 (2017) 241. https://doi.org/10.1617/s11527-017-1108-5.

17. M. Benzerzour, M. Amar, N.-E. Abriak, New experimental approach of the reuse of dredged sediments in a cement matrix by physical and heat treatment, *Constr. Build. Mater.* 140 (2017) 432–444. https://doi.org/10.1016/j.conbuildmat.2017.02.142.

18. M. Benzerzour, W. Maherzi, M. Amar, N.-E. Abriak, D. Damidot, Formulation of mortars based on thermally treated sediments, *J. Mater. Cycles Waste Manag.* 20 (2018) 592–603. https://doi.org/10.1007/s10163-017-0626-0.

19. N. Junakova, J. Junak, Recycling of reservoir sediment material as a binder in concrete, *Procedia Eng.* 180 (2017) 1292–1297. https://doi.org/10.1016/j.proeng. 2017.04.291.

20. N. Junakova, J. Junak, Sustainable use of reservoir sediment through partial application in building material, *Sustainability.* 9 (2017) 852. https://doi.org/10. 3390/su9050852.

21. M. Amar, M. Benzerzour, N.-E. Abriak, Y. Mamindy-Pajany, Study of the pozzolanic activity of a dredged sediment from Dunkirk harbour, *Powder Technol.* 320 (2017) 748–764. https://doi.org/10.1016/j.powtec.2017.07.055.

22. M. Amar, M. Benzerzour, A.E.M. Safhi, N.-E. Abriak, Durability of a cementitious matrix based on treated sediments, *Case Stud. Constr. Mater.* (2018). https://doi.org/10.1016/j.cscm.2018.01.007.

23. M. Amar, M. Benzerzour, N.-E. Abriak, Towards the establishment of formulation laws for sediment-based mortars, *J. Build. Eng.* 16 (2018) 106–117. https:// doi.org/10.1016/j.jobe.2017.12.011.

24. Z. Zhao, M. Benzerzour, N.-E. Abriak, D. Damidot, L. Courard, D. Wang, Use of uncontaminated marine sediments in mortar and concrete by partial substitution of cement, *Cem. Concr. Compos.* 93 (2018) 155–162. https://doi.org/10.1016/ j.cemconcomp.2018.07.010.

25. H. Du, S.D. Pang, Value-added utilization of marine clay as cement replacement for sustainable concrete production, *J. Clean. Prod.* 198 (2018) 867–873. https:// doi.org/10.1016/j.jclepro.2018.07.068.

26. O. Safer, N. Belas, O. Belaribi, K. Belguesmia, N.-E. Bouhamou, A. Mebrouki, Valorization of dredged sediments as a component of vibrated concrete: Durability of these concretes against sulfuric acid attack, *Int. J. Concr. Struct. Mater.* 12 (2018) 44. https://doi.org/10.1186/s40069-018-0270-7.

27. C. Van Bunderen, R. Snellings, L. Vandewalle, Ö. Cizer, Early-age hydration and autogenous deformation of cement paste containing flash calcined dredging sediments, *Constr. Build. Mater.* 200 (2019) 104–115. https://doi.org/10.1016/j.conbuildmat.2018.12.090.

28. C. Van Bunderen, F. Benboudjema, R. Snellings, L. Vandewalle, Ö. Cizer, Experimental analysis and modelling of mechanical properties and shrinkage of concrete recycling flash calcined dredging sediments, *Cem. Concr. Compos.* 115 (2021) 103787. https://doi.org/10.1016/j.cemconcomp.2020.103787.

29. el M. Safhi, M. Benzerzour, P. Rivard, N.-E. Abriak, I. Ennahal, Development of self-compacting mortars based on treated marine sediments, *J. Build. Eng.* 22 (2019) 252–261. https://doi.org/10.1016/j.jobe.2018.12.024.

30. el M. Safhi, M. Benzerzour, P. Rivard, N.-E. Abriak, Feasibility of using marine sediments in SCC pastes as supplementary cementitious materials, *Powder Technol.* (2018). https://doi.org/10.1016/j.powtec.2018.12.060.

31. el M. Safhi, P. Rivard, A. Yahia, M. Benzerzour, K.H. Khayat, Valorization of dredged sediments in self-consolidating concrete: Fresh, hardened, and microstructural properties, *J. Clean. Prod.* 263 (2020) 121472. https://doi.org/10.1016/j.jclepro.2020.121472.

32. el M. Safhi, P. Rivard, A. Yahia, K.H. Khayat, N.-E. Abriak, Durability and transport properties of SCC incorporating dredged sediments, *Constr. Build. Mater.* 288 (2021) 123116. https://doi.org/10.1016/j.conbuildmat.2021.123116.

33. R. Hadj Sadok, N. Belas, M. Tahlaiti, R. Mazouzi, Reusing calcined sediments from Chorfa II dam as partial replacement of cement for sustainable mortar production, *J. Build. Eng.* 40 (2021) 102273. https://doi.org/10.1016/j.jobe.2021.102273.

34. R. Hadj Sadok, W. Maherzi, M. Benzerzour, R. Lord, K. Torrance, A. Zambon, N.-E. Abriak, Mechanical properties and microstructure of low carbon binders manufactured from calcined canal sediments and ground granulated blast furnace slag (GGBS), *Sustainability*. 13 (2021) 9057. https://doi.org/10.3390/su13169057.

35. R. Kou, M.-Z. Guo, L. Han, J.-S. Li, B. Li, H. Chu, L. Jiang, L. Wang, W. Jin, C.S. Poon, Recycling sediment, calcium carbide slag and ground granulated blast-furnace slag into novel and sustainable cementitious binder for production of eco-friendly mortar, *Constr. Build. Mater.* 305 (2021) 124772. https://doi.org/10.1016/j.conbuildmat.2021.124772.

36. H. Mehdizadeh, M.-Z. Guo, T.-C. Ling, Ultra-fine sediment of Changjiang estuary as binder replacement in self-compacting mortar: Rheological, hydration and hardened properties, *J. Build. Eng.* 44 (2021) 103251. https://doi.org/10.1016/j.jobe.2021.103251.

37. R.A. Halassa, M. Bibi, M.-A. Chikouche, Behavior of cementitious materials under the effect of an eco-cement based on dredged sludge, *Ann. Chim.: Sci. Matér.* 45 (2021) 455–465. https://doi.org/10.18280/acsm.450604.

38. A. Bouchikhi, A. el M. Safhi, P. Rivard, R. Snellings, N.E. Abriak, Fluvial sediments as SCMs: Characterization, pozzolanic performance, and optimization of equivalent binder, *J. Mater. Civ. Eng.* (2021). https://doi.org/10.1061/(ASCE)MT.1943-5533.0004071.

39. B. Lothenbach, K. Scrivener, R.D. Hooton, Supplementary cementitious materials, *Cem. Concr. Res.* 41 (2011) 1244–1256. https://doi.org/10.1016/j.cemconres.2010.12.001.

40. D.R. Gaskell, Chapter Fourteen: The determination of phase diagrams for slag systems, in: J.-C. Zhao (Ed.), *Methods for Phase Diagram Determination*, Elsevier Science Ltd, Oxford, 2007: pp. 442–458. https://doi.org/10.1016/B978-008044629-5/50014-8.

41. S. Amziane, Setting time determination of cementitious materials based on measurements of the hydraulic pressure variations, *Cem. Concr. Res.* 36 (2006) 295–304. https://doi.org/10.1016/j.cemconres.2005.06.013.

42. H. Ez-zaki, A. Diouri, Microstructural and physicomechanical properties of mortars-based dredged sediment, *Asian J. Civ. Eng.* 20 (2019) 9–19. https://doi.org/10.1007/s42107-018-0084-6.

43. T. Zdeb, T. Tracz, M. Adamczyk, Effect of the amount of river sediment on the basic properties of cement mortars, *MATEC Web Conf.* 322 (2020) 01032. https://doi.org/10.1051/matecconf/202032201032.

44. E. Berodier, K. Scrivener, Understanding the filler effect on the nucleation and growth of C-S-H, *J. Am. Ceram. Soc.* 97 (2014) 3764–3773. https://doi.org/10.1111/jace.13177.

45. T. Matschei, B. Lothenbach, F.P. Glasser, The role of calcium carbonate in cement hydration, *Cem. Concr. Res.* 37 (2007) 551–558. https://doi.org/10.1016/j.cemconres.2006.10.013.

46. M. Balonis, F.P. Glasser, The density of cement phases, *Cem. Concr. Res.* 39 (2009) 733–739. https://doi.org/10.1016/j.cemconres.2009.06.005.

47. H. Ez-zaki, A. Diouri, S. Kamali-Bernard, Transport properties of blended cement based on dredged sediment and shells, *Adv. Mater. Lett.* 8 (2017) 1–0.

48. H. Zhou, W. Zhang, H. Wei, T. Liu, D. Zou, H. Guo, A novel approach for recycling engineering sediment waste as sustainable supplementary cementitious materials, *Resour. Conserv. Recycl.* 167 (2021) 105435. https://doi.org/10.1016/j.resconrec.2021.105435.

49. Arrêté du 15 février 2016 relatif aux installations de stockage de déchets de sédiments, n.d. https://www.legifrance.gouv.fr/eli/arrete/2016/2/15/DEVP1519170A/jo/texte (accessed July 24, 2020).

50. Arrêté du 12 décembre 2014 relatif aux conditions d'admission des déchets inertes dans les installations relevant des rubriques 2515, 2516, 2517 et dans les installations de stockage de déchets inertes relevant de la rubrique 2760 de la nomenclature des installations classées, n.d. https://www.legifrance.gouv.fr/eli/arrete/2014/12/12/DEVP1412523A/jo/texte (accessed July 24, 2020).

51. F. Avet, R. Snellings, A. Alujas Diaz, M. Ben Haha, K. Scrivener, Development of a new rapid, relevant and reliable (R3) test method to evaluate the pozzolanic reactivity of calcined kaolinitic clays, *Cem. Concr. Res.* 85 (2016) 1–11. https://doi.org/10.1016/j.cemconres.2016.02.015.

52. G. Aouad, A. Laboudigue, N. Gineys, N.E. Abriak, Dredged sediments used as novel supply of raw material to produce Portland cement clinker, *Cem. Concr. Compos.* 34 (2012) 788–793. https://doi.org/10.1016/j.cemconcomp.2012.02.008.

53. A. Faure, A. Smith, C. Coudray, B. Anger, H. Colina, I. Moulin, F. Thery, Ability of two dam fine-grained sediments to be used in cement industry as raw material for clinker production and as pozzolanic additional constituent of portland-composite cement, *Waste Biomass Valorization.* 8 (2017) 2141–2163. https://doi.org/10.1007/s12649-017-9870-8.

54. A. Rehmat, A. Lee, A. Goyal, M. Mensinger, Construction-grade cement production from contaminated sediments using Cement-LockTM technology, (1999).

55. J.L. Dalton, K.H. Gardner, T.P. Seager, M.L. Weimer, J.C.M. Spear, B.J. Magee, Properties of Portland cement made from contaminated sediments, *Resour. Conserv. Recycl.* 41 (2004) 227–241. https://doi.org/10.1016/j.resconrec.2003.10.003.

56. I. Anger, J.-P. Moulin, F.T. Commene, D. Levacher, Fine-grained reservoir sediments: An interesting alternative raw material for Portland cement clinker production, *Eur. J. Environ. Civ. Eng.* 23 (2019) 957–970. https://doi.org/10.1080/19648189.2017.1327890.

57. A. Faure, C. Coudray, B. Anger, I. Moulin, H. Colina, L. Izoret, F. Théry, A. Smith, Beneficial reuse of dam fine sediments as clinker raw material, *Constr. Build. Mater.* 218 (2019) 365–384. https://doi.org/10.1016/j.conbuildmat.2019.05.047.

58. C. Chu, J. Kleib, M. Amar, M. Benzerzour, N.-E. Abriak, Recycling of dredged sediment as a raw material for the manufacture of Portland cement: Numerical modeling of the hydration of synthesized cement using the CEMHYD3D code, *J. Build. Eng.* (2021) 103871. https://doi.org/10.1016/j.jobe.2021.103871.

59. H.F. Taylor, *Cement Chemistry*, Thomas Telford London, 1997.

60. C. Ferone, F. Colangelo, R. Cioffi, F. Montagnaro, L. Santoro, Use of reservoir clay sediments as raw material for geopolymer binders, *Adv. Appl. Ceram.* 112 (2013) 184–189. https://doi.org/10.1179/1743676112Y.0000000064.

61. C. Ferone, B. Liguori, I. Capasso, F. Colangelo, R. Cioffi, E. Cappelletto, R. Di Maggio, Thermally treated clay sediments as geopolymer source material, *Appl. Clay Sci.* 107 (2015) 195–204. https://doi.org/10.1016/j.clay.2015.01.027.

62. K. Komnitsas, Co-valorization of marine sediments and construction & demolition wastes through alkali activation, *J. Environ. Chem. Eng.* 4 (2016) 4661–4669. https://doi.org/10.1016/j.jece.2016.11.003.

63. H. Slimanou, K. Bouguermouh, N. Bouzidi, Synthesis of geopolymers based on dredged sediment in calcined and uncalcined states, *Mater. Lett.* 251 (2019) 188–191. https://doi.org/10.1016/j.matlet.2019.05.070.

64. A. Bouchikhi, M. Benzerzour, N.E. Abriak, W. Maherzi, Y.M. Pajany, Waste glass reuse in geopolymer binder prepared with metakaolin, *Acad. J. Civ. Eng.* 37 (2019) 539–544. https://doi.org/10.26168/icbbm2019.78.

65. S. Hosseini, N.A. Brake, M. Nikookar, Ö. Günaydın-Şen, H.A. Snyder, Enhanced strength and microstructure of dredged clay sediment-fly ash geopolymer by mechanochemical activation, *Constr. Build. Mater.* 301 (2021) 123984. https://doi.org/10.1016/j.conbuildmat.2021.123984.

66. R. Karam, M. Paris, D. Deneele, T. Wattez, M. Cyr, D. Bulteel, Effect of sediment incorporation on the reactivity of alkali-activated GGBFS systems, *Mater. Struct.* 54 (2021) 118. https://doi.org/10.1617/s11527-021-01720-y.

67. ICR Research, *The Global Cement Report^{TM}*, 14th Edition, 2021.

68. AFNOR, NF EN 15167-1: Ground granulated blast furnace slag for use in concrete, mortar and grout - Part 1: Definitions, specifications and conformity criteria, (2006).

69. AFNOR, NF EN 13263-1+A1: Silica fume for concrete - Part 1: Definitions, requirements and conformity criteria, (2009).

70. AFNOR, NF EN 450-1: Fly ash for concrete - Part 1: Definition, specifications and conformity criteria, (2012).

71. C09 Committee, ASTM C1866: Standard specification for ground-glass pozzolan for use in concrete, *ASTM International*, n.d. https://doi.org/10.1520/C1866_C1866M-20.

Index

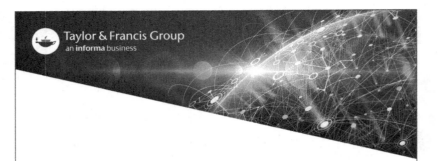

Printed in the United States
by Baker & Taylor Publisher Services